Tunable Materials with Applications in Antennas and Microwaves

Synthesis Lectures on Antennas

Editor
Constantine A. Balanis, *Arizona State University*

Synthesis Lectures on Antennas will publish 50- to 100-page publications on topics that include both classic and advanced antenna configurations. Each lecture covers, for that topic, the fundamental principles in a unified manner, develops underlying concepts needed for sequential material, and progresses to the more advanced designs. State-of-the-art advances made in antennas are also included. Computer software, when appropriate and available, is included for computation, visualization and design. The authors selected to write the lectures are leading experts on the subject who have extensive background in the theory, design and measurement of antenna characteristics. The series is designed to meet the demands of 21st century technology and its advancements on antenna analysis, design and measurements for engineers, scientists, technologists and engineering managers in the fields of wireless communication, radiation, propagation, communication, navigation, radar, RF systems, remote sensing, and radio astronomy who require a better understanding of the underlying concepts, designs, advancements and applications of antennas.

Tunable Materials with Applications in Antennas and Microwaves
John N. Sahalos and George A. Kyriacou
2019

Analysis and Design of Transmitarray Antennas
Ahmed H. Abdelrahman, Fan Yang, Atef Z. Elsherbeni, and Payam Nayeri
2017

Design of Reconfigurable Antennas Using Graph Models
Joseph Costantine, Youssef Tawk, and Christos G. Christodoulou
2013

Meta-Smith Charts and Their Potential Applications
Danai Torrungrueng
2010

Generalized Transmission Line Method to Study the Far-zone Radiation of Antennas under a Multilayer Structure
Xuan Hui Wu, Ahmed A. Kishk, and Allen W. Glisson
2008

© Springer Nature Switzerland AG 2022

Reprint of original edition © Morgan & Claypool 2019

Tunable Materials with Applications in Antennas and Microwaves

John N. Sahalos and George A. Kyriacou

ISBN: 978-3-031-00414-8   paperback
ISBN: 978-3-031-01542-7   ebook
ISBN: 978-3-031-00019-5   hardcover

DOI 10.1007/978-3-031-01542-7

A Publication in the Springer series
SYNTHESIS LECTURES ON ANTENNAS

Lecture #13
Series Editor: Constantine A. Balanis, *Arizona State University*
Series ISSN
Print 1932-6076   Electronic 1932-6084

Tunable Materials with Applications in Antennas and Microwaves

John N. Sahalos
Aristotle University of Thessaloniki, Greece
and University of Nicosia, Cyprus

George A. Kyriacou
Democritus University of Thrace, Xanthi, Greece

SYNTHESIS LECTURES ON ANTENNAS #13

ABSTRACT

Tunable Materials with Applications in Antennas and Microwaves is a stimulating topic in these modern times. With the explosion of the new generation of the wireless world, greater emphasis than ever before is being placed on the analysis and applications of modern materials. This book describes the characteristics of *Ferrites and Ferroelectrics* and introduces the reader to *Multiferroics*.

1. Represents, in a simple manner, the solid state physics and explains the permittivity and permeability tensor characteristics for the tunable materials of infinite and finite dimensions.

2. Gives the applications of tunable materials in resonators, filters, microstrips, striplines, antennas, phase shifters, capacitors, varactors, and frequency selective surfaces.

3. Describes in detail the mathematical analysis for spin and magnetostatic waves for infinite medium, thin slab films, and finite circular discs. The analysis contains original work, which the reader may extend in the future.

4. Provides multiferroics, which are ferrite and ferroelectric composites. Multiferroics are very promising tunable materials which are believed will offer many applications in the near future.

5. Contains the planar transmission lines with analytic formulas for multilayer microstrips, transmission lines, and waveguides with isotropic as well as anisotropic dielectric and magnetic materials. Also, gives the formulas to analyze the layered category of transmission lines with multiferroics.

This book is intended for antenna and microwave engineers as well as for graduate students of Materials Science and Engineering, Electrical & Computer Engineering, and Physics Departments.

KEYWORDS

antennas, microwaves, tunable materials, ferrites, ferroelectrics, multiferroics, transmission lines

Contents

List of Figures

List of Tables

Preface

Tunable Materials with Applications in Antennas and Microwaves is written to meet the needs of antenna and microwave engineers as well as of graduate students of Electrical & Computer Engineering and Physics Departments. The objective of the book is to describe the characteristics of ferrites and ferroelectrics, which are the most common and widely known tunable materials, as well as to introduce the reader to multiferroics. Several applications, especially in modern wireless communication systems, can be found with the above materials.

Essentially, the book represents in a simple manner the solid state physics of tunable materials and explains their permittivity and permeability tensor characteristics. The tensors are given for materials with infinite and finite dimensions.

The five chapters of the book address the following topics:

Chapter 1 is an introductory chapter that contains the basics of tunable materials and the evolution of ferrites and ferroelectrics. The applications of the above materials in resonators, filters, microstrips, striplines, antennas, phase shifters, capacitors, varactors, and frequency selective surfaces are discussed.

Chapter 2 gives the characteristics and the constitutive parameters of ferrites and ferroelectrics. It starts from different types of magnetic materials, magnetic moments, structure of certain ferrites, and the MMICs. It explains the magnetization equation and gives the permeability tensor in different coordinate systems. Also, it discusses the dielectric properties, the polarizability, and the losses of ferrites. For the ferroelectric materials, the chapter represents their properties, the permittivity tensor, their dielectric response, the hysteresis loop, the losses, as well as the crystal structures of certain ferroelectrics.

Chapter 3 presents finite ferrites samples including their spin and magnetostatic modes. Mathematical analysis is given in details for spin and magnetostatic waves for infinite medium, thin slab films, and finite circular discs. This chapter contains original work, which the reader may extend in the future.

Chapter 4 gives multiferroics. These are very promising tunable materials with ferrite and ferroelectric composites. We give certain properties, relations, and interesting characteristics. It is believed that multiferroics will be materials with many applications in the near future.

Chapter 5 contains the planar transmission lines, which are the absolute media in the antenna and microwave applications. In the chapter, we give analytic formulas for multilayer microstrips, transmission lines, and waveguides with isotropic as well as anisotropic dielectric and magnetic materials.

John N. Sahalos and George A. Kyriacou
August 2019

Acknowledgments

We would like to acknowledge the friendship, constructive criticism, encouragement, and patience of our editor, Prof. C. Balanis, who was waiting for the completion of this book for a long time. We also express our gratitude to Dr. Lena Gialabouki and Dr. Dimitrios Babas, who have read the manuscript, proposed improvements, and detected many wording problems. Last but not least, we would like to thank our wives Angela Sahalos and Zoe Kyriacou for their love, support, and patience during the writing of this book.

John N. Sahalos and George A. Kyriacou
August 2019

CHAPTER 1

Ferrites and Ferroelectrics

1.1   TUNABLE ELECTROMAGNETIC MATERIALS

The potential usefulness of tunable electromagnetic materials has been recognized for at least 40 years [1, 2]. Since those early times, ferroelectric and ferrimagnetic materials (ferrites) in their bulk form have been exploited in the implementation of tunable microwave devices. A variety of famous devices, like circulators, phase shifters, and YIG (yttrium iron garnet) tuned oscillators, contain magnetized bulk ferrites as their main part. Bulk ferroelectrics have not yet found their way into widespread microwave applications. This is because ferroelectrics have higher losses at room temperature than ferrites. They also require large bias voltage. Nevertheless, electrical control of ferroelectrics is considered easier than that of ferrites. In the last decade, microwave applications of ferroelectrics have begun to emerge. This is mainly due to material technology advancement, which tends to enhance ferroelectric properties.

The driving force behind the recent revived interest in ferrites and ferroelectrics is their potential for substantial miniaturization of the components and systems in microwave (mw) and millimeter wave (mmw) regions. In particular, the evolution of thin and thick film technology in ferrite/ferroelectric materials enables the integration of the above components with microelectronic circuits. This is also accompanied by cost reduction. Cost, in turn, has a significant impact on the growth of commercial wireless and defense-military systems. In order to realize the importance of miniaturization, just recall that this was the key to the extraordinary evolution of computer hardware and the whole information technology accompanying it. Similarly, the miniaturization of radio frequency electronics underpins the dramatic progress of wireless-mobile communication systems. In turn, miniaturization of mw-mmw circuits is expected to carry the burden of the ever-increasing demand for bandwidth. This is important because "broadband services for all," especially in the fifth generation (5G), of wireless communications, are now in fashion and in great demand.

A question to be expected is how ferrites/ferroelectrics and, in general, tunable materials are involved in the miniaturization game. Moreover, it is extremely useful to know how important tunability in the advancement of mw and mmw electronics is. One of the key features in the miniaturization of the above circuits is the reduction of their guided wavelength (λ_g). This, in turn, depends on the index of refraction (n) as $\lambda_g \propto 1/n$; $n = \sqrt{\mu_r \varepsilon_r}$. Thus, a material with a high dielectric constant ε_r and/or high relative permeability μ_r serves this purpose. Ferrites and composite-ferroelectrics offer high and tunable values of μ_r and ε_r, respectively.

In trying to learn from low frequency electronics, we see that miniaturization is attributed to the employment of the appropriate materials for various electronic devices. These devices are compatible with integrated circuits technology. In this framework, the size of the circuit components is considered similar, in the range of micrometers (μm) or even sub-micrometers. This, in turn, requires the materials to be in the form of films [3].

Recent developments in film processing techniques enable the deposition of materials either in layer by layer or with a gradient in concentration. These techniques also offer precise control over thickness as well as stoichiometry and grain size [3]. However, numerous undesirable effects are observed during the processing of composite and multilayer films. Especially in ferroelectrics, we have [2, 3] the following:

- Ferroelectric thin films are often more lossy than bulk materials of similar composition.

- Ferroelectric thin films often exhibit lower dielectric constant and lower tunability than bulk material of similar composition.

In the following sections, we will elaborate on the necessity of using ferrite/ferroelectric films. But first, an overview of the historical evolution of ferrites will be given.

1.2   OVERVIEW OF THE EVOLUTION OF FERRITES

From a historical point of view, the usefulness of ferrites as tunable materials was realized as early as 1930–1950. Especially after the 1949 publication of Polder's theory [4], the ferrimagnetic resonance influence on magnetic permeability was established as the basis for understanding their microwave properties. Tellegen [5] introduced "gyrator" as a circuit element, while Hogan [6, 7] demonstrated the Faraday rotation in ferrites and proposed microwave components based on it. Many researchers, in turn, proposed a plethora of microwave devices, including non-reciprocal phase shifters and circulators. Bosma [8] and Davies and Cohen [9], paved the way to the design of planar-printed ferrite devices. In [6, 8], a theoretical design of symmetrical junction stripline circulators was given. It was found that the characterization of a ferrite device as reciprocal or non-reciprocal depends, in general, upon its symmetry [9, 10].

It is important to understand the electromagnetic phenomena observed within magnetized ferrites. One of the following effects [1] can explain the behavior of microwave devices that involve ferrites:

- Faraday rotation: When a transverse electromagnetic (TEM) wave propagates in a ferrite along the direction of magnetization, then a rotation of its plane of polarization is observed.

- Ferrimagnetic resonance: A strong absorption occurs when an elliptically polarized radio frequency (RF) magnetic field is applied perpendicularly to the direction of magnetization.

- Field displacement: In a microwave structure partially loaded with ferrite, an increase or decrease in field intensity may occur in the ferrite region due to the displacement of the field distribution transversely to the direction of propagation.

- Nonlinear effects: When a high power RF signal is applied to a ferrite, a nonlinear behavior is observed. This provides unique frequency selective properties and frequency doubling.

- Spin waves [11–16]: When a ferrite is subjected to high RF power levels, the thermal agitation of adjacent magnetic moments may yield spatial variation of magnetization, which may be identified as "spin waves." These spin waves have short wavelengths and may travel at any angle with respect to the direction of magnetization. When their wavelength is comparable to the ferrite dimensions, they are called "magnetostatic waves" (MSW). It should be pointed out that spin waves might also occur at low levels of RF power when the distribution of the external RF magnetic field is non-uniform. Spatial field non-uniformity can be created at the interface of different materials or on the surface of a ferrite. This is, in turn, of particular interest in ferrite films.

1.3   FERROELECTRICS

Tunability in ferroelectrics is based on the variation control of the dielectric constant (ε_r) through a DC-biasing electric field provided by a corresponding DC voltage. Usually, the required biasing voltage is very high, up to tens of kilovolts, (restricted to about 15 KV). However, their DC-biasing consumption is negligible and their variation (switching) speed is of the order of $1\ \mu$sec. Moreover, the microwave elements have relatively high Q of the order of 500 [17–19].

    In principle, any form of ferroelectric materials like bulk ceramics, bulk single crystal, thick films, and thin films can be used as microwave tunable components, and each one of them is accompanied by advantages and disadvantages. In the following sections, several applications will be summarized and their theoretical description along with design guidelines will be given.

1.3.1   BULK FERROELECTRICS

1. Tunable Filters and Tunable Dielectric Resonators
The dimensions of tunable filters and resonators are proportional to the guided wavelength λ_g, which is inversely proportional to the square root of ε_r. For TEM waves, it is $\lambda_g = \lambda_0/\sqrt{\varepsilon_r}$, while for quasi TEM, it is $\lambda_g = \lambda_0/\sqrt{\varepsilon_{reff}}$. ε_{reff} is the effective dielectric constant, which is $\propto \varepsilon_r$. Thus, the dimension dependence is $l \propto 1/\sqrt{\varepsilon_r}$, but this is not always such a simple expression.

2. Tunable Microstrip or Stripline Structures
In tunable microstrip or stripline structures, the bulk ceramic or single crystal ferroelectric is used as a dielectric substrate. Besides the required high bias voltage, the high ε_r value, which is field dependent, yields significant variation in the characteristic impedance. This imposes the additional requirement of an adaptive impedance matching of 50 Ω.

3. Lens Antennas with Beam Steering

Lens antennas are usually implemented as a stack of electroplated ferroelectric slabs [20], used to apply the DC-control voltages across each slab. An appropriate tuning of these DC voltages yields the desired directive beam pointing-steering.

4. Parallel Plate Tunable Capacitors or Varactors

The tunable dielectric constant ε_r of ferroelectrics yields the desired variation in capacitance. Even though their losses are lower than those of semiconductor varactors, the high bias voltage limits their applicabilities.

1.3.2   THIN FILM FERROELECTRICS

1. Varactors

Varactors based on ferroelectric films are made either as planar structures (planar capacitors) or as triple-layered "sandwich" structures [2]. For triple-layered structures, a DC voltage of 1–20 volts is sufficient for effective tuning.

2. Microwave Ferroelectric Phase Shifters

The advancement in phased and adaptive antenna arrays stimulated an impressive research effort in electronically controllable planar phase shifters and especially in integrated circuit form. The purpose of a phased array is to steer its beam toward an arbitrary direction, while adaptive arrays steer the nulls of their radiation patterns toward the direction of interfering sources. The emerging technology of multiple access like SDMA (Space Division Multiple Access) and OFDMA (Orthogonal Frequency Division Multiple Access) in communications and the advanced multifunctional radar require highly integrated *modular* RF front ends (above the intermediate, or IF, stage). Within this topology, each antenna element will be fed by a module of transmit-receive (Tx-Rx), which includes microwave phase shifters [19–21]. The phase shifters are used to control the phase of the signal (channel), exciting each element of the array, which, in turn, steers the beam and, in general, modifies the radiation pattern.

The implementation of phase shifters using ferroelectric films enables their integration with the microwave circuits on the same substrate, preferably as monolithic microwave integrated circuits (MMICs). This primarily serves miniaturization since the antenna dimensions are reduced. In addition, their mass and cost decrease substantially. Moreover, these devices are accompanied by the advantages of ferroelectric films, e.g., control with low DC voltage, low DC-power consumption and increased response speed.

Ferroelectric phase shifters are realized either as analog devices controlled by a DC voltage (which can be digitally controlled through a digital to analog converter (DAC)) or as switched digital phase shifters providing a certain fixed phase shift (45°, 90°, 135°, and 180°). The latter are the corresponding analog of latching ferrite phase shifters [22].

3. Microwave Tunable Filters

A variety of microwave applications in communications, signal intelligence, jamming, and electronic counter measure systems (ECM) requires band-pass filters with variable center frequency and variable bandwidth [23, 24]. Usually, in these applications the frequency of the expected signal (to be received) is somewhere within a very wide band (even a few octaves). The octave range of a device is characterized by the expression {$\log_2$ (Higher/Lower Frequency)}. According to the matched filter theory, the signal to noise ratio is optimized when the filter frequency response H(f) has the same amplitude and opposite phase of the signal spectrum S(f), namely H(f)=S*(f). Otherwise, the noise level at the output of the filter (receiver) increases, causing a serious reduction in receiver sensitivity. As an example, recall that a typical wideband receiver-sensor has a sensitivity of the order of −60 dBm at best. This is not the case of a super-heterodyne frequency selective receiver-sensor, which has a typical sensitivity of −110 dBm. Wideband sensors are usually simple, of low cost and can cover a huge wideband. A typical example is the IFM (Instantaneous Frequency Measurement) receiver used in ECM (Electronic Countermesures) systems. In the classical approach, wideband sensors are cascaded with a bank of narrowband filters connected in parallel (RF pre-selected filters). They have successive center frequencies so that these parallel channels cover the whole band. An alternative approach is the use of a phase locked loop (PLL), but this is costly and complicated. The use of adaptive filters, able to tune their center frequency to that of the signal and to control their bandwidth toward approaching the matched filter, is found to offer an attractive solution. A vast research effort has been devoted to this approach, employing different tunable components. Such components are varactor diodes [25, 26], microwave transistors and, in particular, ferrite loaded microstrip structures [27, 28], ferrite films [28], superconducting films [29], ferroelectric films [30], and Ferrite/Ferroelectric multilayer film structures.

Focusing on the ferroelectric film implementation of tunable filters, we can see that their main advantage is the excellent integration in monolithic microwave chips (miniaturization). However, their high losses still constitute a major problem. For example, in a typical three-pole filter the insertion loss is of the order of 10 dB [28].

4. Microwave Tunable Amplifiers and Oscillators

The ferrite tuned and, in particular, the YIG resonator tuned oscillators are very well established in the microwave range. VCOs (Voltage Controlled Oscillators) based on other techniques (e.g., varactor tuned) offer a tunability of about an octave and any attempt to increase it results in worse spectral purity. In contrast, the YIG possesses a wide-band tuning with good spectral purity [31]. Two disadvantages of YIG resonator tuned oscillators stem from the required strong biasing DC-magnetic field. The first disadvantage is the slow variation whereas an almost instant change is required. The second one comes from the fact that it is very difficult to provide such a strong DC-magnetic field in a monolithic microwave circuit (MMIC) configuration. Moreover, the tuning speed of YIG resonators is severely limited by magnetic hysteresis while modern systems require a tuning speed of more than 1 GHz/μsec [32]. Ferroelectric tuned oscillators could be used to overcome the above difficulties, but the most promising configuration is that of multilayer ferrite-ferroelectric films. The latter are expected to exploit the well-established tunability and spectral purity properties of the YIG (now in a film form) in conjunction with the convenient and fast changing DC voltage tuning of ferroelectrics in MMIC configurations.

5. Multi–Octave Tunable Amplifiers

Wideband microwave amplifiers are usually realized by employing a balanced topology. When a bandwidth of up to 40% (maximum 50%), referring to their center frequency, is desired, a balanced topology with two hybrid couplers in quadrature or with branch line couplers is used, e.g., [22]. When Lange couplers [33] instead of branch line couplers are used, a bandwidth of the order of more than 2.5 octaves can be easy achieved. In both cases, the center frequency of operation could, in principle, be tuned by using a ferroelectric substrate. The change of its permittivity will be realized with the aid of a DC-voltage bias. Some attempts toward this direction are described in [33–35].

A distributed amplifier constitutes a classical wideband topology with a bandwidth of more than a decade, presenting a good input and output matching [22]. The amplifier cannot provide a very high gain, or a low noise figure and it is larger in size than an amplifier of narrower bandwidth and comparable gain.

6. Tunable Microwave Antennas

The research effort toward phased and adaptive arrays and their applications have already been discussed in the section on tunable phase shifters. Within this approach, the maximum and/or nulls of the radiation pattern are steered with the aid of a tunable phase shifter inserted at the input port of each antenna element, e.g., [21]. An alternative approach is to use the tunable material as a substrate supporting the radiator or just as a tunable inclusion (e.g., post) inserted into a dielectric substrate. The variation in either permeability (ferrite) or permittivity (ferroelectric) or both (e.g., ferrite/ferroelectric layers) yields either resonance frequency agility or a beam steering of the antenna, but it may also serve both purposes. The main advantage of this approach is that it eliminates the requirement of a complex beamforming network (BFN) and is accompanied by substantial cost reduction. Futhermore, it may serve a variety of purposes since these features are provided even by a single element. In particular, when traveling wave antennas or the specialized form of "Leaky Wave Antennas" are employed, very directive radiation systems with exceptional characteristics are realized. Moreover, the introduction of EBG (Electromagnetic Band Gap) techniques in the structure of tunable antennas constitutes a state-of-the art with highly ambitious expectations, e.g., super directive antennas with ultra low side lobes, miniaturization, and integration in microwave circuits.

Research on tunable and/or beam steering based on ferrites started some decades ago. However, the main drawbacks of ferrites, namely the difficulties in integration, in high drive power (for the DC-magnetic field biasing) and in slow tuning have set back their potentially widespread application. The above difficulties may be overcome if ferroelectrics are used instead of ferrites. Of course, we must deal with the very high losses, especially when ferroelectric films are used. In addition, the required bias voltage is too high (few KV) for bulk ferroelectric materials [36]. Once again, the multilayer ferrite/ferroelectric structure is expected to be the most appropriate.

7. Tunable Microwave Resonators

Microwave resonators constitute the basic building block of bandpass and bandstop filters. Moreover, they are critical components of oscillators and of various sensors and measuring devices. Tunable filters and tunable oscillators are, likewise, based on tunable resonators. Thus, the performance of the above filters, oscillators and sensors is primarily dependent on the characteristics of the resonators involved, like their quality factor, tunability, etc. The electronic tunability of these resonators is mostly achieved by varying the permeability and/or the permittivity of the materials comprising the resonator structure. For cavity resonators, this is the material contained in the cavity. For printed resonators, the tunable material is either its substrate or its superstrate or an inclusion within each of them. Futhermore, the tunable material losses (loss tangent) strongly affect the resonator quality factor, which, in turn, affects the performance of the constructed microwave device.

Until recently, a major problem concerning the miniaturization of microwave devices is the inability to integrate resonators and, in turn, filters within a monolithic circuit. This is simply due to the basic physical property that microwave resonators have dimensions of at least a quarter $(\lambda_g/4)$, and usually a half $(\lambda_g/2)$, of the guided wavelength. The use of materials such as ferrites and ferroelectrics, with their high relative permeability and permittivity (μ_r and $\varepsilon_r \gg 1$), serve the miniaturization task since λ_g is inversely proportional to the square root of $(\mu_r \varepsilon_r)$. Moreover, the development of resonators printed on ferrite or ferroelectric multilayer film structures enables the integration of resonators into microwave integrated circuits (MIC) as well as monolithic ones (MMICs). Once again, the most promising structure is that of multilayer ferrite/ferroelectric films [15, 37] and [38].

1.4   FERRITE–FERROELECTRIC FILMS

The most serious problem in ferroelectrics and especially those of the thin film form is their high RF losses (high $\tan \delta$). Ferroelectrics, however, have a strong advantage in tunability, which is achieved by the control of permittivity through a biasing DC-electric field and, in turn, a corresponding DC voltage. On the other hand, the main disadvantage of ferrites is the need for electromagnets that provide the variable DC bias magnetic field. Even though ferrites usually have lower RF losses than ferroelectrics, the electromagnets are usually heavy devices with high DC controlling power consumption. This is also true for the advanced implemented "miniature electromagnets" [12].

Fortunately, a recent development in ferrite-ferroelectric layered structures, which support surface electromagnetic spin waves, is expected to offer a solution that combines the advantages of the two materials [13–15]. A coupling is observed between spin waves and delayed electromagnetic waves. The above cause a hybridization of dispersion curves between those of ferrite spin waves and layered structure electromagnetic waves [14]. In this ferrite-ferroelectric structure, the dispersion characteristics of spin waves have been found to depend on the ferroelectric layer dielectric constant [16]. In turn, the characteristics of spin waves can be tuned by the DC voltage bias of the ferroelectric. It has also been found that losses could be significantly reduced in comparison to those of ferroelectric films, e.g., [17]. To sum up, we can say that ferrite-ferroelectric films seem to be very promising structures offering increased tuning speed, decreased controlling DC power, and size reduction (which serves miniaturization).

1.5   TUNABLE FREQUENCY SELECTIVE SURFACES (FSSS)

Frequency selective surfaces (FSSs) are two-dimensional periodic structures, which provide frequency filtering to the incoming electromagnetic waves [39, 40]. The physical properties of FSSs have evolved from diffraction gratings in optics [39] and have been studied in the microwave domain as well. However, in the last decade, FSSs have experienced exceptional advancements with applications in filters, diplexers, reflector antennas, printed antenna substrates, perfect magnetic ground planes, radar-antenna randomes, and "smart surfaces" in stealth techniques. Typical FSSs employed in previous years were static structures. The emerging tunability and reconfigurability of FSSs are expected to expand the functionality and capability of the devices based on them and, in turn, those of high-frequency communication systems. In particular, high-speed tunability and reconfigurability without compromising filtering performance constitute a fundamental challenge in current FSS research [40, 41].

Typical FSSs comprise periodically arranged metallic patch elements printed on a dielectric substrate. Moreover, dual structures are involved in FSSs, namely apertures or slot elements within a metallic screen [39]. In the above cases, the FSSs exhibit total reflection (patches) or total transmission (apertures) in the neighborhood of the elements' resonance. In general, the FSS response is built up of constructive or destructive (or at an arbitrary phase) interference of the waves reflected (or transmitted) by each individual element. It is, thus, obvious that the FSS response depends on the period (spacing), shape, and orientation of the elements of the periodic array as well as the permittivity and permeability of any material loading. In order to clarify the latter dependence, just pay attention to the statement that "the elements' resonance is the basic mechanism affecting the FSS response." However, it is well known that the substrates' (with or without ground plane) permittivity and permeability along with the shape and dimensions of the printed patches determine their resonance. The same holds for metallic screens with apertures when they are loaded with material layers. It is, then, clear that the use of a tunable material (ferrite, ferroelectric) as a substrate or loading yields tunable FSS. The use of ferrite and/or ferroelectric films seem to be the most appropriate tunable loading materials.

Multilayer FSSs have been employed in [41] in order to optimize the performance of a broadband antenna array. Futhermore, switches have been incorporated in the top layer of patches to make the array reconfigurable. Again, ferrite/ferroelectric layers could be employed within this structure to make it tunable.

To the author's knowledge, FSS tunability through tunable materials is still an open issue and there are a few studies in this direction, e.g., [42, 43].

1.6   REFERENCES

[1] J. Douglas Adam, L. E. Davis, G. F. Dionne, E. F. Schloemann, and S. N. Stitzer, Ferrite devices and materials, *IEEE Transactions on Microwave Theory and Techniques*, vol. MTT-50, no. 3, pp. 721–737, March 2002. DOI: 10.1109/22.989957. 1, 2

[2] A. G. Tagantsev, V. O. Sherman, K. F. Astafiev, J. Venkatesh, and N. Setter, Ferroelectric materials for microwave tunable applications, *Journal of Electroceramics*, vol. 11, pp. 5–66, Springer, 2003. DOI: 10.1023/b:jecr.0000015661.81386.e6. 1, 2, 4

[3] K. P. Jayadevan and T. Y. Tseng, Review composite and multilayer ferroelectric thin films: Processing, properties and applications, *Journal of Material Science: Material in Electronics*, vol. 13, pp. 439–459, Springer, 2002. DOI: 10.1023/A:1016129318548. 2

[4] D. Polder, On the theory of ferromagnetic resonance, *Philosophical Magazine*, vol. 40, p. 99, 1949. DOI: 10.1080/14786444908561215. 2

[5] B. D. H. Tellegen, The gyrator: A new electric network element, *Philips Research Reports*, vol. 3, p. 81, 1948. 2

[6] C. L. Hogan, The microwave gyrator, *Bell System Technical Journal*, vol. 31, p. 1, 1952. 2

[7] C. L. Hogan, The ferromagnetic faraday effect at microwave frequencies and its applications, *Reviews of Modern Physics*, vol. 25, p. 253, 1953. DOI: 10.1103/revmodphys.25.253. 2

[8] H. Bosma, On the principle of stripline circulation, *Proc. Institute of Electrical Engineers*, vol. 109B, suppl. 21, pp. 137–146, January 1962. DOI: 10.1049/pi-b-2.1962.0027. 2

[9] J. B. Davies and P. Cohen, Theoretical design of symmetrical junction stripline circulators, *IEEE Transactions on Microwave Theory and Techniques*, vol. MTT-11, pp. 506–512, June 1963. DOI: 10.1109/tmtt.1963.1125717. 2

[10] V. A. Dimitriyev, Symmetry of microwave devices with gyrotropic media—complete solution and applications, *IEEE Transactions on Microwave Theory and Techniques*, vol. MTT-45, pp. 394–401, March 1997. DOI: 10.1109/22.563338. 2

[11] R. F. Soohoo, *Theory and Application of Ferrites*, Prentice Hall, NJ, 1960. 3

[12] J. W. Judy and R. S. Muller, Batch-fabricated, addressable, magnetically actuated microstructures, *Solid-State Sensor and Actuator Workshop*, pp. 187–190, CA, June 1996. 7

[13] V. E. Demidov, B. A. Kalinikos, S. F. Karmanenko, A. A. Semenov, and P. Edenhofer, Electrical tuning of dispersion characteristics of surface electromagnetic-spin waves propagating in ferrite–ferroelectric layered structures, *IEEE Transactions on Microwave Theory and Techniques*, vol. MTT-51, no. 10, pp. 2090–6, October 2003. DOI: 10.1109/tmtt.2003.817461. 8

[14] V. E. Demidov and B. A. Kalinikos, Dipole-exchange theory of hybrid electromagnetic-spin waves in layered film structures, *Journal of Applied Physics*, vol. 91, no. 12, pp. 10007–10016, June 2002. DOI: 10.1063/1.1475373. 8

[15] V. E. Demidov, P. Edenhofer, and B. A. Kalinikos, Electrically tunable microwave phase shifter based on ferrite–ferroelectric structure, *Electronics letters*, vol. 37, pp. 1155–1156, 2001. DOI: 10.1049/el:20010746. 7, 8

[16] W. J. Kim, W. Chang, S. B. Quadri, H. D. Wu, J. M. Pond, S. W. Kirchoefer, H. S. Newman, D. B. Chrisey, and J. S. Horwitz, Electrically and magnetically tunable microwave devices using (Ba, Sr) $TiO_3/Y_3 Fe_5 O_{12}$ multilayer, *Applied Physics A*, vol. 7, p. 71, 2000. DOI: 10.1007/PL00021083. 3, 8

[17] O. G. Vendik, *Ferroelectrics in MW Technology*, Sovetskoe Radio, 1979. 3, 8

[18] M. J. Lancaster, J. Powell, and A. Porch, Thin-film ferroelectric microwave devices, *Superconductor Science and Technology*, vol. 11, pp. 1323–1334, IOP, 1998. DOI: 10.1088/0953-2048/11/11/021.

[19] S. Gevorgian, O. Tageman, and A. Derneryd, Electronically scanning beam-formers based on ferroelectric technology, *Frequenz*, vol. 59, pp. 40–48, 2005. 3, 4

[20] J. Rao, D. Patel, and V. Krichevsky, Voltage controlled ferroelectric lens phased arrays, *IEEE Transactions on Antennas and Propagation*, vol. AP-47, pp. 458–468, 1999. DOI: 10.1109/aps.1996.549911. 4

[21] D. N. McQuiddy, R. L. Gassner, P. Hull, J. S. Masson, and J. M. Bedinger, Transmit/receive module technology for X-band active array radar, *IEEE Proc.*, vol. 79, pp. 308–340, March 1991. DOI: 10.1109/5.75088. 4, 6

[22] D. M. Pozar, *Microwave Engineering*, 4th ed., Wiley, NJ, 2012. 4, 6

[23] G. Mironenko, A. A. Ivanov, A. A. Semenov, S. S. Karmanenko, A.V. Gordeichuk, and T. Inushima. Tunable microwave resonators and filters based on ferroelectric multislot transmission line, *Ferroelectrics*, vol. 286, pp. 343–352, Taylor & Francis, 2003. DOI: 10.1080/00150190390206545. 5

[24] H. Dayal, Variable bandwidth, wide tunable frequency, voltage tuned filter, *International Journal RF Microwave Computer Aided Engineering*, vol. 14, pp. 65–72, 2004. DOI: 10.1002/mmce.10119. 5

[25] S. R. Chandler, I. C. Hunter, and J. G. Gardiner, Active varactor tunable bandpass filter, *IEEE Microwave and Guided Wave Letters*, vol. 3, no. 3, pp. 70–71, March 1993. DOI: 10.1109/75.205668. 5

[26] A. R. Brown and G. M. Rebeiz, A varactor-tuned RF filter, *IEEE Transactions on Microwave Theory and Techniques*, vol. MTT-48, no. 7, pp. 1157–1160, July 2000. DOI: 10.1109/22.848501. 5

[27] I. Huynen, G. Goglio, D. Vanhoenacker, and A. Vander Vorst, A novel nanostructured microstrip device for tunable stopband filtering applications at microwaves, *IEEE Microwave and Guided Wave Letters*, vol. 9, no. 10, pp. 401–403, October 1999. DOI: 10.1109/75.798029. 5

[28] S. D. Silliman, H. M. Christen, L. A. Knauss, K. S. Harshavardhan M. M. A. El Sabbach, and K. Zaki, Magnetically tunable microwave filters based on YBCO/YIG/GGG heterostructures, *Journal of Electroceramics*, vol. 4, pp. 305–310, Springer, 2000. DOI: 10.1023/A:1009902306806. 5

[29] J. P. Contour, K. Bouzehouane, C. Couvert, D. Crété, E. Jacquet, Y. Lemaitre, J. C. Mage, D. Mansart, and B. Marcilhac, Frequency agile filters based on $SrTiO_3$/$YBa_2Cu_3O_7$ heterostructures grown by PLD on $LaAlO_3$ substrates, *Ferroelectrics*, vol. 286, pp. 23–38, Taylor & Francis, 2003. DOI: 10.1080/00150190390212124. 5

[30] G. Subramanyam, F. Van Keuls, and F. A. Miranda, A K-band tunable microstrip bandpass filter using a thin-film conductor/ferroelectric/dielectric multilayer configuration, *IEEE Microwave and Guided Wave Letters*, vol. 8, no. 2, pp. 78–80, February 1998. DOI: 10.1109/75.658647. 5

[31] L. Divina and Z. Skvor, The distributed oscillator at 4 GHz, *IEEE Transactions on Microwave Theory and Techniques*, vol. MTT-46, no. 12, pp. 2240–2243, December 1998. DOI: 10.1109/22.739204. 5

[32] I. C. Hunter and J. D. Rhodes, Electronically tunable microwave bandpass filters, *IEEE Transactions on Microwave Theory and Techniques*, vol. MTT-30, no. 9, pp. 1355–1360, September 1982. DOI: 10.1109/tmtt.1982.1131260. 6

[33] G. Subramanyam, F. A. Miranda, F. Van Keuls, R. R. Romanofsky, C. L. Canedy, S. Aggarwal, T. Venkatesan, and R. Ramesh, Performance of a K-band voltage-controlled lange coupler using a ferroelectric tunable microstrip configuration, *IEEE Microwave and Guided Wave Letters*, vol. 10, no. 4, pp. 525–530, April 2000. DOI: 10.1109/75.846924. 6

[34] J. E. G. Colom, R. Medina, and R. Rodriguez, Design of tunable balanced amplifier using ferroelectric materials, *Integrated Ferroelectrics*, vol. 56, pp. 1097–1106, December 2003. DOI: 10.1080/10584580390259650.

[35] J. E. G. Colom, R. Medina, and Y. Perez, Simulation of single-stage tunable amplifier using ferroelectric materials, *Integrated Ferroelectrics*, vol. 56, pp. 1131–1140, December 2003. DOI: 10.1080/10584580390259795. 6

[36] Y. Yashchyshyn and J. W. Modelski, Rigorous analysis and investigations of the scan antennas on a ferroelectric substrate, *IEEE Transactions on Microwave Theory and Techniques*, vol. MTT-53, no. 2, pp. 427–438, February 2005. DOI: 10.1109/tmtt.2004.840779. 7

[37] E. J. Yan, C. I. Cheon, High frequency tunable LC devices with ferroelectric/ferromagnetic thin film heterostructure, *Physica Status Solidi*, (b)241, no. 7, pp. 1625–1628, Wiley, 2004. DOI: 10.1002/pssb.200304604. 7

[38] J. Wosik, L. M. Xie, M. Strikovski, J. H. Miller, and P. Przyslypski, Microwave characterization of $Na_0, 67 Sr_0, 33 MnO_3 - x$ thin films of magnetically tunable filters, *Journal of Applied Physics Letters*, vol. 74, pp. 750–752, 2001. DOI: 10.1063/1.123191. 7

[39] R. MiTtra, C. H. Chan, and T. Cwik, Techniques for analyzing frequency selective surfaces—a review, *Proc. IEEE*, vol. 76, pp. 1593–1615, December 1988. DOI: 10.1109/5.16352. 8

[40] J. Zendejas, J. Gianvittorio, B. Yoo, Y. Rahmat Samii, K. Nobe, and J. W. Judy, Micro machined magnetically reconfigurable frequency selective surfaces, *Proc. of Solid State Sensors, Actuators and Microsystems Workshop*, CA, June 2002. 8

[41] Y. E. Erdemli, K. Sertel, R. A. Gilbert, D. E. Wright, and J. L. Volakis, Frequency-selective surfaces to enhance performance of broad-band reconfigurable arrays, *IEEE Transactions on Antennas and Propagation*, vol. AP-50, pp. 1716–1724, December 2002. DOI: 10.1109/tap.2002.807377. 8, 9

[42] E. A. Parker and S. B. Savia, Active frequency selective surfaces with ferroelectric substrates, *IEE Proc. Microwaves, Antennas and Propagation*, vol. 148, pp. 103–106, 2001. DOI: 10.1049/ip-map:20010306. 9

[43] D. Kuylenstierna, A. Vorbiev, G. Subramanyam, and S. Gevorgian, Tunable electromagnetic bandgap structures based on ferroelectric films, *IEEE APS*, pp. 879–882, Columbus, OH, 2003. DOI: 10.1109/aps.2003.1220412. 9

CHAPTER 2

Tunable Materials–Characteristics and Constitutive Parameters

2.1 INTRODUCTION

Tunable materials belong to the family of modern structures that will improve engineering technology. According to WUN (Worldwide Universities Network) [1], "In the future new materials and manufacturing processes will transform our experience of everyday life. Smart structural materials will enhance safety, comfort and fuel efficiency in transportation; new biomimetic materials may enable the replacement of organs or body tissue; new generations of display devices will open up new possibilities in global communications and contribute to a greener environment through reduced energy consumption." The above statement, though general, gives the direction of the evolution of new technologies. Herein, we will focus on high frequency communication technologies and, particularly, on the exploitation of tunable materials for the development of electronically controllable radio frequency (RF) and microwave (mw) components and devices. The advantages provided by tunable materials have already been discussed in Chapter 1. It should be recalled that they serve miniaturization and integration of RF or mw front ends into monolithic circuits and, most importantly, multi-functionality. Tunable RF and mw devices and circuits provide us with the flexibility to adapt to various changes in operating conditions. Such conditions are the operating frequency, the input impedance, and the power level inherent in wireless communications [2]. The availability of components that can be tuned over a broad range enables the operation of communication systems over multiple frequency bands. Tunable materials preserve the highly desired advantages and the characteristics of frequency selective devices. A typical wideband receiver (e.g., one based on crystal detectors) suffers from high noise level and has poor sensitivity (always lower than about −60 dBm). On the other hand, frequency selective receivers (e.g., heterodyne) have excellent sensitivity and low noise level as well as low out-of-band interference. However, they are restricted to a narrowband operation. This problem was solved by mechanically tuning their front-end frequency of operation and by retaining a constant frequency at their intermediate stages. Modern communications and defense systems cannot use the above technique because they require fast frequency agility (or frequency hopping) of the order of 1 GHz/μsec [3].

Devices based on tunable materials can cover this demand and can offer new features like adaptive impedance matching for amplifiers and antennas, e.g., [4]. The tuning of these systems may be implemented via computer, microcontroller (μC), and digital signal processor (DSP) control or via a feedback loop embedded in the system. Moreover, the easily provided power control can be used to compensate for the deleterious effects of aging and temperature sensitivity variations of RF-mw circuits [2]. In addition, power control serves the energy-saving requirements in mobile units.

Besides the above, the driving force behind the revolutionization of high frequency communications is *miniaturization* of RF and mw circuits. Miniaturization is an unbreakable bond with integration into monolithic microwave circuits (MMICs). The concept "system on chip" is very well known and widely accepted. It has been the driving force behind the incredible revolution in microelectronics as well as digital circuits and computers. Furthermore, it has already been tested in RF electronics, where one can see exceptional advancements in mobile communications. Now, it is expected to revolutionize communications and defense systems for higher frequencies (mw; millimeter waves). By integrating the device into a chip, we enable mass production, which, in turn, dramatically lowers the cost, making this technology affordable for everyone. Increased demand, especially in commercial applications, acts as feedback to technology advancements, forming a loop with an imaginable evolution. Ferroelectric films fulfill the need for miniaturization and integration in tunable devices. Even better performance is expected after using ferrite/ferroelectric film structures. Artificial wire media and metamaterials also constitute very promising technologies. Most of these technologies are at the stage of investigation and evaluation.

Some of the primary alternative technologies to implement tunable RF and mw devices are the following [2]:

1. Solid state gallium arsenide field-effect transistors (GaAs-FET) and GaAs varactor diodes. These tend to have low Q (quality factor) due to high losses and cannot handle high power levels.

2. Micro-machined (MEMS)-based devices, whose reliability is still to be tested and determined. Moreover, they often have difficult biasing requirements and stringent packaging needs.

This chapter will first focus on the microwave properties of tunable or agile materials. These will be discussed both in their bulk form, as thick films, and in their most desired form for integration purposes, the thin film form. Bulk ferrites have been extensively analyzed in the literature and a short overview will be presented. In addition, bulk ferroelectrics in conjunction with high temperature superconductors will also be discussed.

2.2   MICROWAVE FERRITES

2.2.1   HISTORICAL EVOLUTION

The discovery of ferrites along with magnetism itself dates back to about 600 BC and is attributed to the Greek philosopher Thales of Miletus [5]. Miletus, the motherland of Thales, is located in Asia Minor. Thales observed permanent magnet properties in the natural mineral magnetite ($Fe^{+2} Fe_2^{+3} O_4^{-2}$), which he named "Magnetes lithos" in Greek, namely magnet rock. Though always known, the importance of ferrites was not recognized until Snoek and his co-workers started working at Philips Laboratories around 1936, e.g., [6]. Snoek sought for "magnetic insulators of high frequency transformer cores, or for the realization of high permeability without eddy current losses" [5]. In addition, Hogan [7] derived the Faraday rotation and its applications, especially the gyrator. At the same time, ferrites were seen as a breakthrough in microwave technology [7]. Due to this, an IRE (Institute of Radio Engineers) Proceedings special issue was devoted to them in October 1956. According to [8], the frequency range in magnetic technology is from 300 MHz to 300 GHz, but mainly up to about 70 GHz is well exploited. Furthermore, present trends in microwave magnetics are toward using frequencies higher than 30 GHz with *miniaturization*. It should be noted that in 1988, the magnet business amounted to some \$30 billion annually, which was greater than that of the semiconductor industry [8].

2.2.2   UNIQUE FERRITE FEATURES

Many applications, especially in mw, are based on the unique features of ferrites. These include non-reciprocity, non-linear gyro-magnetic behavior, resonance, magneto-optics, and presence of magneto-static waves or spin waves (also called "magnons," "polaritons," and "gyromagnetics waves" [8]). The theory on wave propagation to be given in the next chapter will briefly cover these phenomena. Moreover, research into the use of ferrites at higher microwave frequencies (e.g., above 30 GHz) uncovers new phenomena and new mechanisms that could be exploited for the integration of ferrites (in particular, in thin film form) within MMICs. Examples include the *millimeter cyclotron resonances* in HEMTs, (High Electron Mobility Transistors), or *magnetoplasmons* [8, 9], as well as *solitons* [8, 10]. Cyclotron resonances at millimeter wave are observed when a DC magnetic field is applied normally to the GaAs/AlGaAs interface and the thin two-dimensional electron gas of the HEMT structure. This phenomenon may enable the development of a new class of non-reciprocal components readily integrated into the semiconductor structures. Solitons are nonlinear responses of ferrites in the form of non-dispersive short pulses. It has been found that YIG could be the ideal ferrite for these applications.

2.2.3   INTEGRATION OF MICROWAVE MAGNETIC

In the current trends toward integration of microwave magnetic with MMICs, ferrite films are mostly employed. This is because of the unique properties of ferrites. The properties include their wide-band nonreciprocal operation, power limiting ability, time delay characteristics and tunability. An important question to be addressed is whether single-crystal ferrites and polycrystal ferrites, all oriented in the same direction (oriented ferrites) should be integrated. Single-crystals are superior to polycrystals since they allow a larger time delay, while, at the same time, they have lower losses. However, the choice between single-crystal ferrites and oriented ferrites remains an open question [8].

Moreover, the multilayer ferrite film, alternated with a ferroelectric or semiconductor material, is a more promising integration approach. In particular, ferrite-ferroelectric multilayer films based on spin waves render the need for a biasing magnetic field redundant. Another approach that is used is that of dividing MMIC regions into magnetic, dielectric, or semiconductor "islands." These islands derive the area required for each circuit.

Concerning the integration of ferrites with MMICs [8, 11], the progress in thin film biasing magnets through the deposition of high-quality oriented thin film permanent magnets is also very encouraging. Additionally, some applications may render the need for a biasing magnetic field redundant by just exploiting the built-in anisotropy field [8].

Finally, according to [8], successful integration tends to require in RF and mw circuit power levels consistent with a small number of magnetic materials. Moreover, the best approach is to employ single crystals for both semiconductor and RF magnetic circuits. This will reduce the cost associated with single crystal development, while the high fabrication cost may be compensated for by the expected mass production of MMICs. Engineering on the atomic scale for the evolution toward this direction is required. This is important both in terms of design and in terms of production (e.g., fabricate multimagnetic layers and regions even one atomic layer at a time) [8].

2.2.4   BASIC PROPERTIES OF MAGNETIC MATERIALS

The magnetic properties of materials are a consequence of electron motion. This consists of: (i) the orbital motion around the nucleus, and (ii) the electron spin (magnetic moment due to the angular rotation around itself).

Since the electron carries a negative electric charge $(-e)$, its motion creates equivalent current densities in a direction opposite that of the velocity vectors. Both types of motion are rotational (e.g., elliptic); thus, the equivalent currents give rise to effective current loops, which in turn can be accounted for as equivalent magnetic moments.

1. Orbital Magnetic Moments

The effect of orbital magnetic moments is generally insignificant for two reasons. First, electron orbits are randomly oriented, yielding a zero net magnetic moment per unit volume [12]. Second, due to an external magnetic field, the orbital magnetic moments cannot be aligned along that field because they are constrained by the material cohesion forces.

2. Spin Magnetic Moments

Concerning the spin magnetic moments, in most solids, especially in compounds, electron spins occur in pairs with opposite signs. This yields a negligible overall magnetic moment [13]. However, valence electrons and electrons in incompletely filled shells may have unpaired spins, which result in a net magnetic moment. Materials of the transition series of the periodical table (Mg, Mn, Co, Ni, Fe, Cr, Ti, etc.), in particular give rise to a strong overall magnetic moment due to their unpaired electrons of the incomplete shells (especially the 3d sub-shell). More specifically, Hund's rule [12] states that the electron of an incomplete shell will be so oriented that a maximum number of unpaired spins is created. Even in these magnetic materials, electron spins and, consequently, spin moments are randomly oriented so that their net magnetic moment is still negligible. The most important feature that yields magnetic properties is that spin magnetic dipole moments in each atom are free to align themselves along any applied external magnetic field. After the removal of the external magnetic field, adjacent electron spin moments may preserve their alignment due to the existence of exchange forces. The above gives rise to permanent magnetization phenomena, e.g., [12, 13]. The alignment tendency of these elementary magnetic dipoles along the external magnetic field yields an increase in the total magnetic flux density within the material.

2.2.5   ELECTRON MAGNETIC MOMENT—BOHR MAGNETON

The magnetic moment (m_0) of an electron orbiting around the nucleus on a shell of radius r is given in Figure 2.1 [12]. Herein, a normalized m_0 is considered, where free space permeability μ_0 is dropped. Thus, m_0 reads as follows:

$$m_0 = \frac{1}{2} e \cdot r^2 \cdot \omega, \tag{2.1}$$

where $e = 1.602 \times 10^{-19}$ Cb is the electron charge and ω is the rotation circular frequency. The above equation is derived by equating the electrostatic force of attraction, $e^2/(4\pi\varepsilon_0 r^2)$, between the electron and the nucleus with the centripetal force $m_e r \omega^2$. ε_0 is the free space permittivity and $m_e = 9.107 \times 10^{-31}$ Kg is the electron mass. In addition, $\mu_0 = 4\pi \cdot 10^{-7}$ H/m and $\varepsilon_0 = 8.854 \cdot 10^{-12}$ F/m.

The electron's magnetic moment due to its spin can be expressed in terms of its angular momentum. This is obtained from quantum mechanics as [12] $j = \hbar/2 = h/4\pi$, where $\hbar = h/2\pi$ and $h = 6.547 \times 10^{-34}$ J · sec $\approx 4.1148 \times 10^{-15}$ eV · sec is Planck's universal constant. In fact, $j = n \cdot \hbar$, where n is the quantum number. This number takes the values of $n = \pm 1/2$ for an electron spin. Therefore, the magnetic moment of an electron due to its spin, also called "Bohr magneton," (μ_B) is [12–14]:

$$m = \mu_B = \frac{e}{m_e} j = \frac{e}{m_e} \frac{h}{4\pi} = 9.274 \times 10^{-24} \text{ Am}^2. \tag{2.2}$$

Magnetic Moments

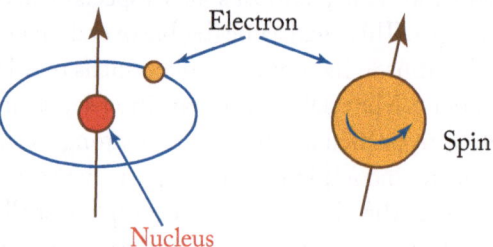

Figure 2.1: An electron orbiting around the nucleus.

The ratio of the magnetic moment m over the electron's angular momentum j is called "gyromagnetic ratio γ." It is:

$$\gamma = \frac{m}{j} = \frac{e}{m_e} = 1.759 \times 10^{11} \ \text{Cb/Kg}. \tag{2.3}$$

In vector notation, $\bar{m}$ and $\bar{j}$ are antiparallel due to the electron's negative charge, namely,

$$\bar{m} = -\gamma \bar{j}. \tag{2.4}$$

The contribution of each electron spin to flux density is equal to:

$$\mu_0 \mu_B = 1.165 \times 10^{-29} \ \text{Wb} \cdot \text{m}. \tag{2.5}$$

As mentioned previously, the contribution of the orbital magnetic moment to the overall magnetic moment of the electron is insignificant. The relative contribution of orbital and spin magnetic moments are accounted for by Landé's g factor:

$$m = gn\mu_B = g\frac{1}{2}\mu_B, \tag{2.6}$$

where $g = 1$ when the moment is due only to orbital motion and $g = 2$ when the moment is due only to spin. For microwave ferrites, the range of g is observed between 1.98 and 2.01. Thus, $g \approx 2$ is a good approximation.

As will be explained below, the magnetic moment of an atom is due to its electrons with unpaired-unbalanced spins. Therefore, its magnitude is equal to the number of unbalanced spins multiplied by the Bohr magneton. Obviously, the magnetic moment of an atom is simply measured in magnetons.

2.2.6   PROPERTIES AND TYPES OF MAGNETIC MATERIALS

Magnetic materials are traditionally categorized in terms of their magnetic properties. These properties are grouped as follows:

1. Diamagnetism

Materials with their entire electron spin balanced owe their weak magnetic properties solely to electron orbital motion (around the nucleus) and are called "***diamagnetic.***" This weak magnetization opposes any applied external magnetic field. Equivalently, it can be represented by very small *negative* susceptibility. The tiny deviation from free space permeability ($\mu_0 = 4\pi \cdot 10^{-7}$ H/m) is typically of the order of -10^{-8} to -10^{-5}. These weak magnetizations are ignored when studying electromagnetic fields in materials and, of course, in the corresponding practical applications (see Figure 2.2).

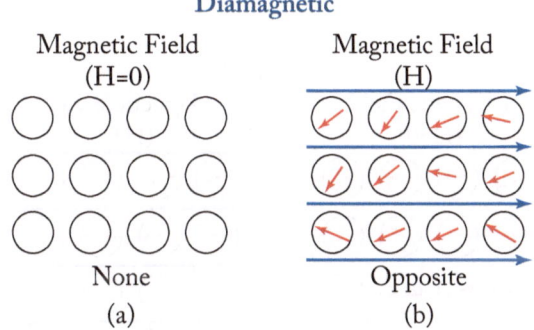

Figure 2.2: Diamagnetic material (a) without and (b) with applied magnetic field.

2. Paramagnetism–Ferromagnetism–Curie Temperature

A material is called ***paramagnetic*** if the electron spin and its orbital motion both contribute to magnetization (see Figure 2.3).

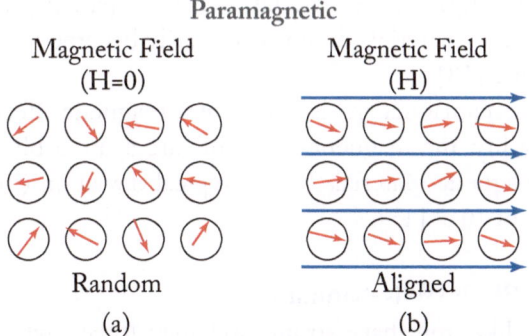

Figure 2.3: Paramagnetic material (a) without and (b) with applied magnetic field.

As mentioned before, when an external magnetic field is applied, the atoms' net moments *tend* to align. A deflection from complete alignment is observed due to the atoms' thermal agility. In some of the important magnetic materials (e.g., Fe and elements of the transition series), the net magnetic moment ($\bar{M}$) of an atom is strong enough, (e.g., due to multiple unbalanced electron spins) to produce an intrinsic-internal magnetic field ($\bar{H}_i$) sufficient to hold the adjacent magnetic dipole aligned, even in the absence of an external field. Thus, it is $\bar{H}_i = K\bar{M}$. The constant K is estimated to be of the order of $K = 1000$ to hold the adjacent atomic moments aligned in parallel. Naturally, the same holds for all adjacent magnetic dipoles; the exerted forces are called "exchange forces" or "exchange interaction" and the phenomenon is called "*ferromagnetism*" (see Figure 2.4).

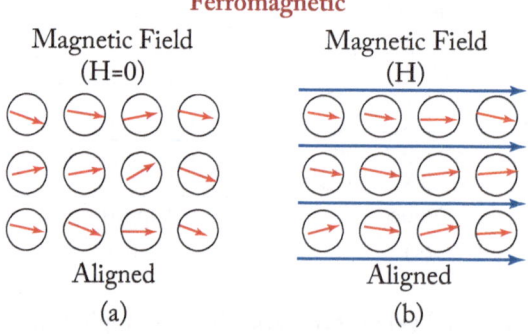

Figure 2.4: Ferromagnetic material (a) without and (b) with external magnetic field.

Note that the thermal vibration of atoms is always present, and above a certain characteristic temperature, called *Curie temperature (Tc)*, this phenomenon becomes dominant for any magnetic material. Materials with strong exchange forces are called *ferromagnetic*. These materials have a Curie temperature higher than ordinary room temperature (17°C or 290 K). Above the Curie temperature, the same materials behave as paramagnetic. However, it should be stressed that paramagnetic materials do not necessarily become ferromagnetic when cooled below a certain temperature [13].

Paramagnetism, like diamagnetism, is a very weak magnetic effect. This happens because, due to atoms' thermal agility, the magnetic moments are again arbitrarily oriented. At room temperature, most paramagnetic materials have a typical deviation (increase) from free space permeability of the order of 10^{-5} [15].

3. Magnetized Domains–Bloch Walls–Saturation

Ferromagnetic materials, like iron, have strong exchange forces, which keep adjacent atoms' magnetic moments aligned in parallel (maximum magnetizations). This happens even in the absence of an external magnetic field, and the phenomenon is called "spontaneous magnetization." If all magnetic moments were aligned in parallel, then the internally stored energy would be maximized. However, the natural tendency of systems is to rearrange their internal structure so that their stored energy is minimized. This, in turn, requires zero overall magnetization. For example, freshly annealed iron specimens or other magnetic materials do not exhibit spontaneous magnetization. These two contradictory requirements, namely the parallel alignment of adjacent magnetic moments and the zero overall magnetization, are fulfilled through the subdivision of the material into small regions called "*domains.*" The magnetic dipole moments within each domain are completely aligned. Therefore, each domain acts as a larger magnetic dipole called "domain magnetic dipole." Domain magnetic dipoles are arbitrarily oriented, and the resulting net magnetization (stored energy minimization) is zero.

When an external magnetic field is applied to a ferromagnetic material, the walls of its domains, also called "Bloch walls," are moved so that the domains can be enlarged. The components of the dipole vectors are in the direction of the applied magnetic field [15]. If the external magnetic field is further increased, the "magnetic dipoles" at non-aligned domains are rotated as a whole toward the direction of the field. This process is non-linear since the dependence of magnetization $\bar{M}$ (magnetic moment per unit volume) and, hence, the total flux density in the material $\bar{B}$ on the applied field intensity $\bar{H}$ is non-linear. Let us assume that a DC external magnetic field $\bar{H}_{DC}$ is applied. As H_{DC} is increased, some domain dipoles are relatively easily turned (aligned). Further increase in H_{DC} causes less domain alignment, leading to a saturation effect. Saturation magnetization M_S is then asymptotically achieved by an ever increasing $\bar{H}_{DC}$.

4. Antiferromagnetism

The parallel alignment of adjacent magnetic moments within a domain is very sensitive to inter-atomic distance in solids [12]. According to Waldron [13], parallel alignment occurs when the ratio of distance d between the nuclei of atoms containing electrons with unpaired spins to radius r of the shell containing the unpaired electrons is a little over 3 ($d/r \approx 3^+$). When the inter-atomic distance d is too large (or $d/r \gg 3$), exchange forces become ineffective and the atomic magnetic moments will not be aligned. Baden Fuller [12] says, "When the inter-atomic spacing d drops below a certain value, the sign of the exchange forces changes and an antiparallel configuration of spin magnetic moments is energetically preferable." The antiparallel magnetic moments cancel each other out; thus, the overall magnetic moment of each domain vanishes and the material appears to be non-magnetic. This phenomenon is called "*antiferromagnetism.*" Waldron [13] regards antiferromagnetism as a special case of ferromagnetism. The difference between ferromagnetism and antiferromagnetism is shown in Figure 2.5.

Pure manganese is an antiferromagnetic material, while its alloys with aluminium (Al) or bismuth (Bi) are ferromagnetics. This occurs because its interatomic distance is modified. The behavior of antiferromagnetics vs. temperature is similar to that of ferromagnetics; in other words, above a certain temperature, called "Néel temperature," the exchange forces can no longer resist the thermal vibration of the lattice and the material becomes paramagnetic.

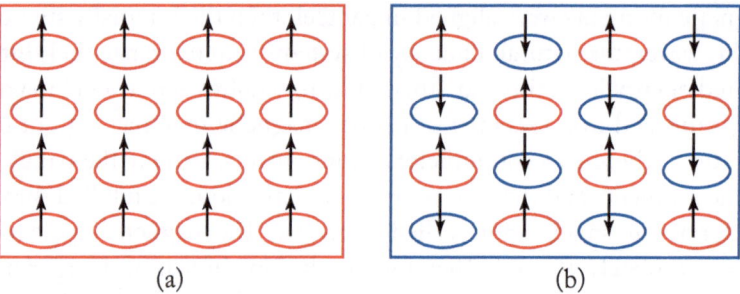

(a)   (b)

Figure 2.5: The difference between (a) ferromagnetism and (b) antiferromagnetism.

5. Ferrimagnetism—Antiferromagnetism

Some materials similar to antiferromagnetics have a crystal structure comprising two interlaced lattices of ions. The magnetic moments of atoms within one lattice are aligned in parallel. The same holds for the other lattice, but the magnetic moments of the two lattices are antiparallel. However, the number of magnetic atoms of the two lattices is, in general, non-equal, resulting in a non-vanishing net magnetization. These materials are called *"ferrimagnetics"* or just *"ferrites."* Like ferromagnetics and antiferromagnetics, the magnetic moments of ferrites form certain domains. They exhibit similar behavior as temperature increases and become paramagnetic above the Curie temperature. The saturation magnetization of ferrites is considerably lower than that of ferromagnetics since their magnetic moments are partially antiparallel. According to Waldron [13], the special case in which the antiparallel magnetic moments cancel each other out is what is called *"antiferromagnetism."* An important characteristic of ferrites (which is also shared by many antiferromagnetics) is that they are composed of oxides of some magnetic metals. Each metallic ion is surrounded by oxygen ions. In some ferrites, magnetic ions may be partially substituted by non-magnetic ions. Their oxide composition provides extremely high resistivity (typically $\rho = 10^6$ to $10^8 \ \Omega \cdot \text{cm}$), which, in turn, gives rise to their unique microwave behavior [16]. Thus, high frequency electromagnetic waves may enter the ferrite, propagate within it, and pass through without excessive reflection or attenuation, respectively. Simultaneously, ferrites provide the required magnetic properties. In contrast, ferromagnetic materials have extremely low resistivity. For example, iron has $\rho_{Fe} \approx 10^{-5} \ \Omega \cdot \text{cm}$. Moreover, the oxide composition of ferrites is also accompanied by a relatively high dielectric constant at microwave frequencies, which ranges from $\varepsilon_r = 10$ to 20. In Table 2.1, the magnetic characteristics of these materials are given.

Table 2.1: Magnetic characteristics of materials

Magmetism	Susceptibility	Magnetic Properties	Material/ Susceptibility
Diamagnetism	Negative and small	Zero magnetic moments	Cu/-0.77×10exp(-6) Au/-2.74×10exp(-6)
Paramagnetism	Positive and small	Randomly oriented magnetic moments	Pt/21.04×10exp(-6) Mn/-66.10×10exp(-6)
Ferromagnetism	Positive and large	Parallel aligned magnetic moments	Fe/~100,000
Antiferromagnetism	Positive and small	Mixed parallel and antiparallel aligned magnetic moments	Cr/3.60×10exp(-6)
Ferrimagnetism	Positive and large function of applied magnetic field	Antiparrallel aligned magnetic moments	Ba/~3 ferrite

2.3 FERRIMAGNETICS: FERRITE MATERIALS AND MAGNETIC GARNETS

2.3.1 SPINNEL FERRITES

1. Chemical Composition

The first developed ferrites (around and after 1956) were ceramic-like, polycrystalline magnetic oxides with an iodestone-spinnel structure [5, 11]. Their generic chemical formula is $M^{+2} Y_2^{+3} O_4^{-2}$, where Y is usually trivalent iron ($M^{+2} Fe_2^{+3} O_4^{-2}$). M is a divalent metallic ion such as $M^{+2} =$ nickel (Ni), cobalt (Co), manganese (Mn), magnesium (Mg), cadmium (Cd), chromium (Cr), or, in general, a metal of the transition series of the periodical table [12]. It is also possible for M^{+2} to represent two types of ions, one of which is monovalent (e.g., lithium Li^{+1}) and the other trivalent (e.g., Fe^{+3}) [13]. A useful spinnel ferrite is $(Li^{+1} Fe^{+3})_{1/2} Fe_2^{+3} O_4^{-2}$.

2. Crystal Structure

An example of a spinnel crystal structure [5, 13] is given in Figure 2.6. A point of interest is that there are two types of lattice sites (A and B) within the spinnel crystal to be occupied by the metallic ions (M^{+2} or Fe^{+3}). A sites, called tetrahedral, have four oxygen neighbors and B sites, called octahedral, have six oxygen neighbors. Within each unit cell, there are 4 A sites, 16 B sites and 32 oxygen ions. The magnetic moments of ions, occupying A and B sites, are aligned by exchange energy in such a way as to be anti-parallel. In this case, the net spontaneous

$$(A^{2+})^{IV} (B^{3+})^{VI}_2 O_4$$

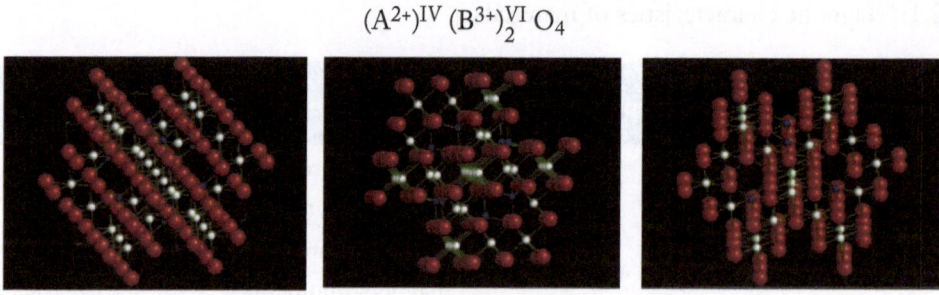

Red (Oxygen), Blue (IV sites), White (VI sites)

Figure 2.6: The unit cell of a spinel crystal structure at various angles.

magnetization (M) is the difference of the magnetization of the corresponding sub lattices M_A and M_B [5], namely, $M = M_B - M_A$. This partial cancellation results in a lower net magnetization and is called "ferrimagnetism." In the case in which all magnetic moments are parallel, the phenomenon of ferromagnetism occurs.

3. Controlling Saturation Magnetization

The saturation magnetization (M_S) of spinels may be controlled by the judicious choice of A and B site ions. To increase M_S, one could increase M_B and decrease M_A. This can be achieved by a partial substitution of the M^{+2} ion with a divalent nonmagnetic ion (e.g., Zn^{+2}), which has a strong preference for A site occupancy. For example, $NiFe_2 O_4$ and $MnFe_2 O_4$ become $Ni_X Zn_{1-X} Fe_2 O_4$ and $Mn_X Zn_{1-X} Fe_2 O_4$.

The above approach causes a reduction in Curie temperature T_C because the alignment of the magnetic moments at A and B sites is coupled through exchange of energy [5]. Thus, a reduction in M_B also results in a decrease in M_A and, in turn, a decrease in M_S as the temperature increases. Recall that for temperatures equal or greater than the Curie temperature, the magnetic moments' alignment is destroyed, driving saturation magnetization to zero. Moreover, the partial replacement of trivalent iron ions (Fe^{+3}) with non-magnetic trivalent ions (e.g., Al^{+3}), which exhibit a strong preference for occupying B sites (e.g., $NiAl_X Fe_{2-X} O_4$), results in a greater reduction in M_B than in M_A and, in turn, a reduction in M_S accompanied by a decrease in T_C.

To conclude this section, spinel ferrites were the first to be invented and used in microwave applications. These first developed ferrites were found to have unacceptably high absorption losses in the microwave range. However, these are now improved. The most commonly used spinel ferrites are: nickel ferrite ($NiFe_2 O_4$), lithium ferrite (($Li^{+1} Fe^{+3}$)$_{1/2}$ $Fe_2 O_4$), and magnesium-manganese ferrite ($Mn_X Mg_Y Fe_Z O_4$), where $x + y + z = 3$ [12]. Their basic characteristics, such as their saturation magnetization M_S and Curie temperature T_C, along with a recommendation for their use, can be seen in Table 2.2. Complete data, including both magnetic and dielectric losses, can be found in the manufacturers' data sheet, e.g., Trans-Tech [17].

Table 2.2: Characteristics of currently used spinnel ferrites [12]

Formula	Curie Temperature	Saturation Ms	Applications
$Mn_xMg_yFe_zO_4$ $0.10 < x < 0.15$ $0.4 < y < 0.5$ $0.7 < x < 0.9$ $x + y + z = 3$	~300°C Temperature sensitive	0.20 - 0.24 Tesla	7–15 GHz - Without wide temperture variations - Low loss and RF power - Rectangular hysteresis loop, for latching applications
$NiFe_2O_4$ $Ni_xZn_{1-x}Fe_2O_4$ $NiAl_xFe_{2-x}O_4$	~570°C	~0.32 Tesla increased by Z_{inc} (Z_n) substitution up to 0.50 Tesla Decreased by Al^{+3} substitution ~0.05 Tesla	- Up to mm-waves (≈30 GHz) using Z_{inc} substitution - High RF power capability
$(Li^{+1}Fe^{+3})_{1/2}Fe_2O_4$	~645°C	~0.36–0.50 Tesla by Zinc substitution Can be decreased by Titanium Ti^{+3} substitution for iron Fe^{+3}	Low loss in MW Rectangular hysteresis loop for latching applications

An extensive, detailed analysis of spinnel ferrite materials can be found in the textbooks of Baden Fuller [12] and Waldron [13], particularly in the references of the original works cited therein.

Note 1: Since all iron ions are trivalent, there is no loss mechanism related to electron transfer between divalent and trivalent iron ions.

Note 2: High saturation magnetization yields gyromagnetic resonances at higher microwave frequencies and, thus, makes the ferrite appropriate for higher microwave bands.

2.3.2   HEXAGONAL FERRITES OR HEXAFERRITES–PERMANENT MAGNETIC FERRITES

Self-Biased Ferrites   The distinguishing feature of hexagonal ferrites is their high residual magnetization, which enables the development of "self biased" microwave devices in the mw frequency range (e.g., above 30 GHz). It is expected that hexagonal ferrites can be used for non-reciprocal functions at frequencies up to or higher than 100 GHz. For this purpose, their large built-in anisotropy field is exploited. However, most of them have a uniaxial permeability tensor [18, 19]; that is, hexagonal ferrites are mostly used in the mw range in order to eliminate the need for a very high external DC bias magnetic field.

Chemical Composition   Hexagonal ferrites are, in general, complex crystalline structures with complicated chemical formulas, which have been given code letters. One large family based on barium (Ba^{+2}) is $(Ba^{+2}O)_X (Me^{+2}O)_Y (Fe_2O_3)_Z$. Me^{+2} is a divalent metallic ion from the first series of transition elements or a combination of elements with valence equal to two [12]. The most widely known representatives are $BaFe_{12}O_{19}$, with the code letter M (also known as BaM) and $Me_2BaFe_{16}O_{27}$, with the code letter W. Another widely used hexagonal ferrite is based on strontium: this is $SrFe_{12}O_{19}$, known as SrM.

Saturation Magnetization   The saturation magnetization and the anisotropy field of BaM and SrM can be reduced by partial substitution of cations. In addition, for the W type $Ni_2BaFe_{16}O_{27}$, a partial substitution of nickel (Ni) by cobalt (Co) lowers its saturation magnetization, while aluminum's (Al^{+3}) substitution for iron (Fe^{+3}) increases both its saturation magnetization and its anisotropy.

   It is important to note that in order to achieve low microwave losses, the ferrite should be produced with all iron ions in trivalent state so that losses related to electron transfer between divalent and trivalent iron ions can be prevented.

2.3.3   MAGNETIC GARNETS

It was as early as 1956 that Bertaut and his coworkers [20] synthesized the first magnetic rare-earth iron garnet in Grenoble. An avalanche of works followed and by the end of the decade, magnetic garnets, including YIG as the most important along with spinel ferrites, had become available to microwave engineers [19].

Crystal Structure   The garnet crystal structure is very complicated. Each unit cell is composed of formula units as $A_3 B_2 C_3 O_{12}$, where A, B, and C are trivalent metallic cations [12]. Within this crystal structure, metal cations are surrounded by oxygen anions in tetrahedron, octahedron, and dodecahedron coordination, occupied correspondingly by C, B, and A ions. The YIG formula is $Y_3Fe_5O_{12}$ or $Y^{+3} \rightarrow A$ and $Fe^{+3} \rightarrow B$ plus C. Thus, Yttrium occupies the dodecahedron and iron occupies both the tetrahedron and octahedron sites in the crystal structure. An example of the crystal structure of the garnet is shown in Figure 2.7.

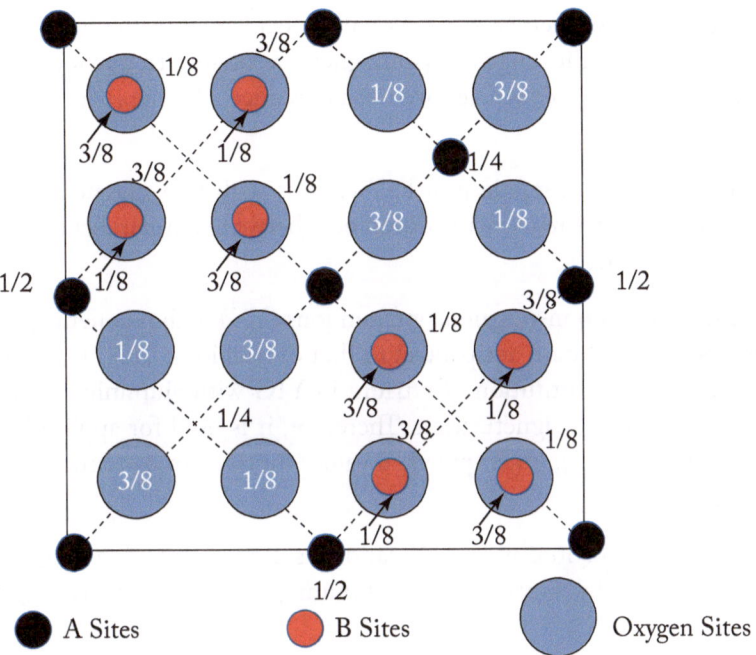

Figure 2.7: Crystal structure of magnetic garnets on the plane $z = 0$.

The fractional numbers show the heights obove the plane. Figure 2.7 shows only the lower half part. The upper half is rotated by 90°.

The magnetic moments of the magnetic cations on different lattice sites are aligned in two antiparallel directions. Thus, garnets are of a ferrimagnetic nature. In YIG, the four iron cations are balanced (yttrium is a non-magnetic ion), resulting in one unbalanced trivalent iron cation (five Bohr magnetons) per unit cell, which gives YIG magnetization. This phenomenon clearly explains why magnetic garnets have a saturation magnetization lower than that of ferrites. In garnets, only one iron ion per cell contributes to net magnetization.

Microwave Losses   It is important to observe that in garnets, all metallic cations are trivalent and there are no positions for divalent ones in the crystal lattice. Thus, losses due to electron exchange between divalent and trivalent ions are clearly avoided, which ensures low losses at microwave frequencies. Moreover, Fe ions provide weak coupling between the excitation of the spin lattice (called "magnons") and the excitation of the crystal lattice (phonons), constituting a negligible direct relaxation loss mechanism [11]. To minimize microwave losses, the garnet should also be prepared as a mono-crystal so that anisotropy effects will not introduce a broadening of gyromagnetic resonance (losses) [11]. Furthermore, the surface of the crystal should be free of damage, and minimal losses are achieved with YIG single crystal spheres.

Controlling Saturation Magnetization   As will be explained in the next section, saturation magnetization (M_s) defines the range of gyromagnetic resonance frequency. Therefore, in order to use YIG at low microwave frequencies, we need to reduce M_s. The reduction is achieved with the following techniques [11]:

1. Partial substitution of magnetic iron Fe^{+3} cations with non-magnetic aluminum cations Al^{+3} on tetrahedral sites reduces the differences between antiparallel magnetizations. This yields a decrease in M_s.

2. Partial substitution of non-magnetic yttrium ions (Y^{+3}) with magnetic gadolinium (Ga^{+3}) ions presents magnetization antiparallel to that of iron ions. This, again, results in M_s reduction. The partial substitution of yttrium in YIG with aluminium (Al) or gadolinium yields lower saturation magnetization. Therefore, it is used for applications at lower microwave frequencies [12]. Exact practical values for M_s can be found in commercial data sheets, e.g., [20, 21].

In general, partial substitution of YIG cations is used with a wide range of metallic ones, e.g., aluminum (Al^{+3}), gadolinium (Ga^{+3}), holmium (Ho^{+3}), calcium (Ca^{+3}), and vanadium (V^{+3}).

Garnets vs. Spinel Ferrites   The saturation magnetization of YIG at room temperature is about $M_s = 0.18$ tesla, and its Curie temperature is $T_C = 286°C$. Furthermore, YIG characteristics have smaller variations with temperature as compared to those of magnesium-manganese ferrites. This, along with its quite lower losses, makes YIG preferable for low power and for control-tunable applications above 3.3 GHz (stemming from its saturation magnetization). Moreover, magnetic garnets have a rectangular hysteresis loop, making them appropriate for latching control applications.

It should be noted at this point that ferrites still preserve the advantage of a much higher saturation magnetization. This allows their operation in the mw frequency bands. In particular, hexagonal ferrites with their self-biasing capability (permanent magnets) can be used at frequencies of up to 100 GHz without the need for an otherwise extremely high required DC biasing magnetic field. As will also be discussed in later sections, this is a desired operation around the gyromagnetic resonance. That is, in turn, calculated as 28 GHz/tesla. In other words, operation at 56 GHz requires a 2 tesla biasing field, which requires bulky electromagnets or superconducting magnets [12].

2.4 FERRITE FILMS

2.4.1 GARNET MONOCRYSTALLIZED FILMS

The trend toward miniaturization of microwave circuits requires tunable materials compatible with MMICs, which are accompanied by lower cost [22]. The ultimate goal is the combination of a ferrite and a semiconductor to achieve a system-on-chip. According to Glass [11], until recently, there was only limited research in ferrite films, focusing mainly on magnetic recording. Epitaxial YIG and calcium vanadium indium (CaVIn) films attracted particular attention in bubble memory technology and in isolators in optical fiber systems [19]. The above can be readily employed for the development of integrated tunable microwave circuits and magnetostatic devices. The key to this direction is high quality *single-crystal films*, which provide minimal microwave losses. Single-crystal YIG films are routinely prepared through the liquid phase epitaxy process (LPE). This is a variation of high-temperature solution growth, usually used for bulk crystals.

Thus, narrowband tunable microwave circuits like filters and YIG-tuned oscillators can be readily integrated into MMICs. It is also noted [21, 22] that their propagation characteristics are influenced by the uniformity of the YIG crystals. This affects the operation of some microwave circuits, like magnetostatic wave devices (MSW). YIG films present an additional advantage since their uniformity is higher than that of bulk single crystals. A very promising configuration is composed of YIG films deposited on a gadolinium (Gd) gallium (Ga) garnet substrate, which is also known as GGG substrate. An example of the integration of a YIG film into a GGG is shown in Figure 2.8.

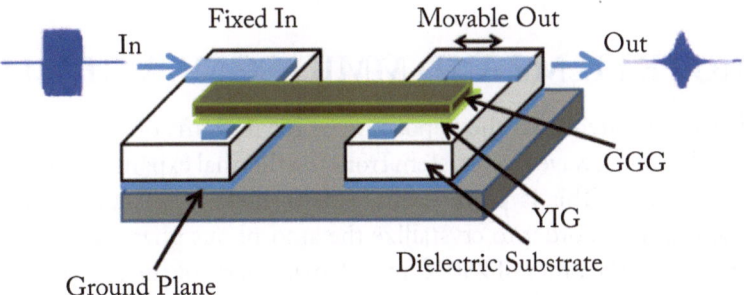

Figure 2.8: Integration of YIG films into a GGG substrate.

Spinel and Hexagonal Monocrystallized Films   Besides the high quality of YIG, its saturation magnetization is lower than that of spinel and hexagonal ferrites. Thus, the development of spinel and hexagonal ferrite films is inevitable for mw integrated circuits. Note, also, that integration into the mw band is even more critical due to the required high accuracy of manufacture (very small wavelength). Therefore, it is highly convenient to exploit the high accuracy available in MMIC fabrication processes. In general, the development of single crystal films of spinel and hexagonal ferrites is possible, but it involves many problems in practical applications (see [11] and the references therein). One of the most serious problems pertains to the surface quality of spinel ferrite films. When slow cooling grows films, they tend to crystallize in the form of an octahedron, yielding a "rough surface" with hillocks and terraces as compared to the smooth surface of crystallized garnet films. This phenomenon results in microwave losses (increased linewidth). A lot of effort is put into magnesium spinel ferrite, $Mg(InGa)_2O_4$, and, in particular, lithium ferrite films. The latter in principle provides low microwave losses, comparable to YIG. It retaines the highest saturation magnetization and the highest Curie temperature.

Self-Biasing   The primary interest in the effort to develop single-crystal hexagonal ferrites is stimulated by their self-biasing capability (due to the large built-in magnetic field) and their high saturation magnetization. These features make them very attractive for mw applications. BaM hexaferrite ($BaFe_{12}O_{19}$), in particular, has attracted the attention of LPE (liquid phase epitaxy) film growers (see [11] and references therein). However, more research effort is required toward their practical exploitation.

2.4.2   POLYCRYSTALLINE FERRITE FILMS

Even though monocrystalline ferrite films constitute the main target, current research has achieved the fabrication of mw and mmw components such as circulators, isolators, and phase shifters. This employs thick film polycrystalline ferrite technology (see [11, 19] and original references therein). Note that films with thickness of 1 μm are considered to be thin films, while films with thickness of the order of 30–35 μm are regarded as thick films.

2.5   FERRITE FILMS AND MMIC COMBATIBILITY

The main problems associated with the deposition of ferrite films on semiconductor substrate, aiming at a monolithic microwave circuit, stem from the thermal expansion mismatches between the film and the substrate. This is particularly evident in the high temperatures required for post-deposition annealing in order to crystallize the amorphous phases of ferrites [11, 19]. For example, temperatures of the order of 250°C may damage parts of the MMIC circuit. However, alternative configurations may solve this compatibility problem. According to Glass [11], it is possible to use oxide substrates, which are appropriate for the preparation of epitaxial GaAs films and for the deposition of ferrite films, even single-crystal ones. In this manner, a single MMIC may be composed of "active semiconductor" and "ferrite film" islands. These regions could be incorporated into optical or acoustical circuits. A significant research effort toward this direction has been activated; the GGG and Si are tried as common substrates (see references in [11]).

Another compatibility problem of ferrite circuits and MMICs stems from the external DC biasing magnetic field. This gets more serious as the frequency of operation increases, especially for mm wave bands. This occurs because increasing frequency entails a higher biasing field to maintain appropriate gyromagnetic resonance (28 GHz/tesla). If we follow the classical approach, the biasing magnets will add unaffordable size and weight. However, there are three alternative solutions:

1. The exploitation of internally built magnetic fields in hexaferrite films for self-biasing.

2. The use of thin-film permanent magnets, e.g., [23, 24].

3. The employment of multilayer ferrite/ferroelectric films based on spin waves, where only a DC voltage biasing of the ferroelectric is needed.

The third approach will be analyzed in an upcoming section.

2.6   FERRITE CONSTITUTIVE RELATIONS

The relation between magnetic flux density $\bar{B}$ and magnetic field intensity $\bar{H}$ in a tensor algebra form [22–29], due to anisotropy, was first set by Polder [28, 29]. The permeability tensor is known as Polder's tensor. A detailed analysis can be found in several textbooks, e.g., [12, 14].

Let us consider a static magnetic field $\bar{H}_0$ (DC bias) applied on a ferrite material (see Figure 2.9).

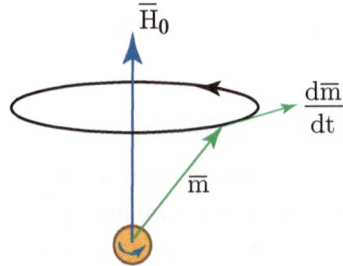

Figure 2.9: Magnetization precession with no damping.

The torque exerted on an unbalanced spin electron will be

$$\bar{T} = \bar{m} \times \bar{B}_0 = \mu_0 \bar{m} \times \bar{H}_0 = -\mu_0 \gamma \bar{j} \times \bar{H}_0. \tag{2.7}$$

This also causes a change in the angular momentum ($\bar{j} = -\bar{m}/\gamma$) with a time rate equal to the torque:

$$\frac{d\bar{j}}{dt} = \bar{T} \qquad \Leftrightarrow \qquad -\frac{1}{\gamma}\frac{d\bar{m}}{dt} = \mu_0 \bar{m} \times \bar{H}_0. \tag{2.8}$$

Rewriting (2.8) yields the well-known equation for the motion of a magnetic dipole moment:

$$\frac{d\bar{m}}{dt} = -\mu_0 \gamma \bar{m} \times \bar{H}_0. \tag{2.9}$$

Assuming, for example, the DC biasing field applied along the z-axis $\bar{H}_0 = H_0\hat{z}$ and a rectangular coordinate system, the solution of (2.9) can be written as:

$$m_x = A\cos(\omega_0 t), \quad m_y = A\sin(\omega_0 t), \quad m_z = C = \text{const.} \qquad (2.10)$$

$\omega_0 = \gamma\mu_0 H_0$ is called "Larmor or free precession frequency."

By observing Eq. (2.10) and Figure 2.9, it becomes obvious that the magnetic moment performs a uniform precession around an axis parallel to the DC biasing magnetic field ($\hat{z}$-axis). However, in Eq. (2.9), losses are ignored, while, in reality, significant loss mechanisms exist. This damping causes the magnetic moment to have a precession which follows a spiral orbit, (see Figure 2.10), until $\bar{m}$ is aligned with the DC biasing field $\bar{H}_0$ (e.g., along $\hat{z}$).

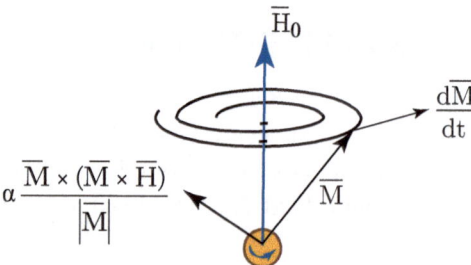

Figure 2.10: Magnetization precession with damping.

2.6.1   MAGNETIZATION EQUATIONS

The magnetic moments of unbalanced electron spins form the atom's magnetic moment. All of them are aligned in a certain direction (the direction of easy magnetization in crystalline ferrites) within a domain of the ferrite, thus having the same vector magnetic moment $\bar{m}$. Therefore, a unit volume with N unbalanced electrons will behave as a larger dipole with a magnetic moment $\bar{M} = N\bar{m}$ (or the magnetization vector). The application of a static biasing field yields the magnetization equation in the same way as for a single electron. In turn, when losses are ignored, the magnetization equation can be written as:

$$\frac{d\bar{M}}{dt} = -\mu_0\gamma\bar{M} \times \bar{H}, \qquad (2.11)$$

where $\bar{H}$ is the internally applied magnetic field, which can be the vector sum of the DC bias field $\bar{H}_0$ and an AC magnetic field $\bar{H}_{AC}$ (e.g., RF-mw). Note that when a ferrite specimen is brought into a biasing field $\bar{H}_0$, its value is, in general, different within the ferrite, depending on the boundary conditions.

Losses can be accounted for in the magnetization equation (see Figure 2.10) by adding a damping term. This was first proposed by Landau and Lifshitz [30], and modified later (see [11,

12] and the references therein):

$$\frac{d\bar{M}}{dt} = -\mu_0\gamma\,\bar{M} \times \bar{H} - \frac{\alpha}{|\bar{M}|}\bar{M} \times \frac{d\bar{M}}{dt}, \qquad (2.12)$$

where α is the damping factor.

According to Waldron [13], "a sufficient understanding of loss mechanisms has not been achieved to provide a physical basis of a theory of losses." Equation (2.12), which is the one commonly used, along with the following Eq. (2.13), which is the Bloch and Bloembergen approach, constitute just two possible formal treatments.

$$\frac{d\bar{M}}{dt} = -\mu_0\gamma\,\bar{M} \times \bar{H} - \bar{M}/\tau, \qquad (2.13)$$

where τ represents a relaxation time of the phenomenon whereby the system loses energy.

However, both the damping factor α and the relaxation time τ can be estimated from the measurements of microwave losses [13]. This definition will be given in the next section.

Once again, it is important to keep in mind that the above magnetization equations are accurately valid under the saturation ($\bar{H}_{DC}$ large enough to align all magnetic moments in its direction) and small-signal ($\bar{H}_{AC} \ll \bar{H}_{DC}$) conditions [13]. For large microwave AC fields, this theory breaks down and non-linear phenomena become significant.

An alternative and more general expression, cited by Morgenthaler [31], and extracted from the original references therein, reads the magnetization equation as follows:

$$\frac{d\bar{M}}{dt} = -\gamma\mu_0\bar{M} \times \left(\bar{H} + \bar{H}^m\right), \qquad (2.14)$$

where $\bar{H}^m$ denotes the contribution of the material to the internal magnetic field consisting of anisotropy $\bar{H}^{anis}$, quantum mechanical exchange $\bar{H}^{ex}$, magnetoelastic $\bar{H}^{me}$, and dissipation components $\bar{H}^{loss}$:

$$\bar{H}^m = \bar{H}^{anis} + \bar{H}^{ex} + \bar{H}^{me} + \bar{H}^{loss}. \qquad (2.15)$$

Note that losses in (2.14) are accounted for in theory by adding a term to the material field. The quantum mechanical exchange is the coupling, which tries to enforce exact parallel alignment of adjacent magnetic moments within a domain. The anisotropy field is related to the phenomenon of spontaneous (in the absence of an external field) magnetization within a domain, where magnetic moments are aligned in a direction of easy magnetization (multiple such axes may exist), creating an internal field [12].

For most practical microwave applications, the ferrite is operated in its saturation region by applying the appropriate DC biasing (H_{DC}) in order to minimize hysteresis loss (as will be explained next). In view of this approach, the "small signal" condition $H_{AC} \ll H_{DC}$ holds. For most practical applications, the above effects of Eq. (2.15) can be ignored and only the loss effects can be preserved. That is, we can use Eqs. (2.12), (2.13), or (2.14) with $\bar{H}^m = \bar{H}^{loss}$.

2.6.2   PERMEABILITY TENSOR

The aim of this section is to relate the magnetic induction vector $(\bar{B})$ and the magnetization $(\bar{M})$ of a magnetically saturated ferrite medium to its magnetic field intensity $(\bar{H})$. In other words, the aim is to extract the corresponding constitutive relations under the small signal conditions. For this purpose, the magnetization Eqs. (2.12) and (2.13) for the lossless and lossy cases, respectively, will be solved. Consider a DC biasing magnetic field in the $\hat{z}$-direction, $\bar{H}_0 = H_0\hat{z}$. Its intensity is appropriate to drive the ferrite to saturation $(M \approx M_S)$. This is usually achieved when $H_0 \geq 4\pi M_S$. Moreover, a small, compared to the DC bias, $|\bar{H}| \ll H_0$, time-harmonic magnetic field $\bar{H} \propto e^{j\omega t}$ (with coordinates H_x, H_y, H_z) is applied. The total magnetic field within the ferrite is $\bar{H}_t = H_0\hat{z} + \bar{H}$. Following the magnetization equation, this gives rise to a total internal magnetization $\bar{M}_t = M_S\hat{z} + \bar{M}$ (note that $\bar{M}$ is the AC quantity caused by the AC field $\bar{H}$). In turn, Eq. (2.11) reads as follows:

$$\frac{d\bar{M}}{dt} = -\gamma\mu_0\bar{M}_t \times \bar{H}_t. \tag{2.16}$$

Solving (2.16) in a Cartesian coordinate system and exploiting the small signal approximation $(H_0 + H_z \approx H_0, M_S + M_z \approx M_S)$ we get the tensor magnetic susceptibility $[\boldsymbol{\mathcal{X}}_m] = [\boldsymbol{\mathcal{X}}]$ relating the AC quantities $\bar{M}$ and $\bar{H}$ (e.g., [12, 14]) as:

$$\bar{M} = [\boldsymbol{\mathcal{X}}]\bar{H} = \begin{bmatrix} \boldsymbol{\mathcal{X}}_{xx} & \boldsymbol{\mathcal{X}}_{xy} & 0 \\ \boldsymbol{\mathcal{X}}_{yx} & \boldsymbol{\mathcal{X}}_{yy} & 0 \\ 0 & 0 & 0 \end{bmatrix}\bar{H} \tag{2.17}$$

where

$$\boldsymbol{\mathcal{X}}_{xx} = \boldsymbol{\mathcal{X}}_{yy} = \frac{\omega_0\omega_m}{\omega_0^2 - \omega^2}, \qquad \boldsymbol{\mathcal{X}}_{xy} = -\boldsymbol{\mathcal{X}}_{yx} = \frac{j\omega\omega_m}{\omega_0^2 - \omega_m^2} \tag{2.18}$$

and

$$\omega_0 = \mu_0\gamma H_0, \qquad \omega_m = \mu_0\gamma M_S. \tag{2.19}$$

The Polder tensor permeability is readily obtained by substituting $\bar{M}$ into the expression of the total AC induction vector within the ferrite as follows:

$$\bar{B} = \mu_0\left(\bar{M} + \bar{H}\right) = [\mu]\bar{H}. \tag{2.20}$$

The $\hat{z}$-bias tensor is:

$$[\mu] = \mu_0\{[U] + [\boldsymbol{\mathcal{X}}]\} = \begin{bmatrix} \mu & jk & 0 \\ -jk & \mu & 0 \\ 0 & 0 & \mu_0 \end{bmatrix} \tag{2.21}$$

where

$$\mu = \mu_0 \left(1 + \mathcal{X}_{xx}\right) = \mu_0 \left(1 + \frac{\omega_0 \omega_m}{\omega_0^2 - \omega^2}\right) \tag{2.22a}$$

$$k = -j\mu_0 \mathcal{X}_{xy} = \mu_0 \frac{\omega \omega_m}{\omega_0^2 - \omega^2}. \tag{2.22b}$$

Tunability   The most important observation in Eqs. (2.21) and (2.22) is that the permeability tensor elements depend on $\omega_0 = \mu_0 \gamma H_0$. This enables the tunability of microwave circuits involving ferrites through the variation of the DC biasing magnetic field H_0, e.g., by using an electromagnet.

Note that when the direction of the bias is reversed (from $H_0\hat{z}$ to $-H_0\hat{z}$), M_S will change sign. In turn, the sign of k will be reversed, but that of μ will remain unchanged. The permeability tensor for $\hat{x}$-bias or $\hat{y}$-bias can be easily obtained from (2.21) by a circular alternation of the subscripts of its elements (see Pozar [14]).

Concerning the units of magnetic quantities, the centimeter-gram-second (CGS) system is traditionally used for most practical tasks [14]. The corresponding relations with the SI system are for the magnetic induction, that is, one tesla, $1\,\mathrm{T} = 10^4$ gauss, field intensity $1\,\mathrm{A/m} = 4\pi \cdot 10^{-3}$ oersted, and $\mu_0 = 1$ gauss/oersted, resulting in the same numerical values for B and H as $B_0(\text{gauss}) = H_0$ (oersted) for a non-magnetic material ($\mu = \mu_0$). The characteristic frequencies of ferrites are calculated as:

CGS units:

$$f_0 = \frac{\omega_0}{2\pi} = \frac{\gamma}{2\pi}\mu_0 H_0 = 2.8\,\mathrm{MHz/oersted} \cdot H_0 \text{ (oersted)}$$

$$\tag{2.23a}$$

$$f_m = \frac{\omega_m}{2\pi} = \frac{\gamma}{2\pi}\mu_0 M_S = 2.8\,\mathrm{MHz/oersted} \cdot 4\pi M_S \text{ (gauss)}$$

SI units:

$$f_0 = 28\,\mathrm{GHz/tesla} \cdot (\mu_0 H_0) \text{ tesla}$$

$$\tag{2.23b}$$

$$f_m = 28\,\mathrm{GHz/tesla} \cdot (\mu_0 M_S) \text{ tesla}.$$

2.6.3   AXIAL MAGNETIZATION IN CYLINDRICAL COORDINATES

The only condition needed for the permeability tensor to be valid is the rotational symmetry along the axis of the static-biasing magnetic field [12]. Therefore, for a DC biasing magnetic field along the $\hat{z}$-axis in the circular cylindrical coordinates (ρ, φ, z), the permeability tensor can be obtained through the Cartesian-to-cylindrical coordinates transformation [32]:

$$\begin{bmatrix} B_\rho \\ B_\phi \\ B_z \end{bmatrix} = \begin{bmatrix} \mu & -jk & 0 \\ jk & \mu & 0 \\ 0 & 0 & \mu_0 \end{bmatrix} \begin{bmatrix} H_\rho \\ H_\phi \\ H_z \end{bmatrix}. \tag{2.24}$$

The expressions for μ and k are given by Eqs. (2.22a) and (2.22b). The similarity of Eq. (2.24) to (2.21) is obvious.

2.6.4   CIRCUMFERENTIAL MAGNETIZATION IN CYLINDRICAL COORDINATES

Very interesting phenomena, enabling corresponding applications, occur when the ferrite is biased transversely to the direction of propagation (usually it is considered along the z-axis). A well-known phenomenon is "birefringence," e.g., Pozar [14], where the propagation of two waves is observed. These are: (1) the "*ordinary wave*," which is unaffected by magnetization, and (2) the "*extraordinary wave*," which depends on magnetization. For antennas printed on ferrite substrate, if the magnetization is perpendicular to the printed structure, the resonant frequency is tunable while the radiation pattern is unaffected [33]. In contrast, if the DC bias is parallel to the substrate but perpendicular to the direction of the propagating wave below the radiator, then a tunable radiated beam steering is enabled, e.g., [34].

In cylindrical coordinates (ρ, φ, z), a very practical way to implement transverse magnetization is the application of a DC biasing magnetic field in the azimuth direction $\bar{H}_0 = H_0 \hat{\phi}$. In this case, the permeability tensor is quite similar to that of $\hat{y}$ biasing in Cartesian coordinates, e.g., Baden Fuller [12]:

$$\begin{bmatrix} B_\rho \\ B_\phi \\ B_z \end{bmatrix} = \begin{bmatrix} \mu & 0 & jk \\ 0 & \mu_0 & 0 \\ -jk & 0 & \mu \end{bmatrix} \begin{bmatrix} H_\rho \\ H_\phi \\ H_z \end{bmatrix}. \tag{2.25}$$

2.6.5   MAGNETIZATION AT AN ARBITRARY DIRECTION IN CARTESIAN COORDINATES

Suppose that we have a DC biasing magnetic field $\bar{H}_0$ (Figure 2.11) oriented along some arbitrary direction (φ_0, θ_0), that is,

$$\bar{H}_0 = H_0 \left\{ \sin\theta_0 \cos\varphi_0 \hat{x} + \sin\theta_0 \sin\varphi_0 \hat{y} + \cos\theta_0 \hat{z} \right\}. \tag{2.26}$$

The permeability tensor was originally given by Tyras [35], and repeated by Rahmat Samii [36], as well as by Hsia [18]. Furthermore, for the case in which $\theta_0 = 90°$, which means that $\bar{H}_0$ is parallel to the x–y plane, the permeability tensor has been given by Hsia and Alexopoulos [34]. Finally, for $\bar{H}_0$ parallel to the x–z plane or $\varphi_0 = 0°$, the permeability tensor, appears in [12].

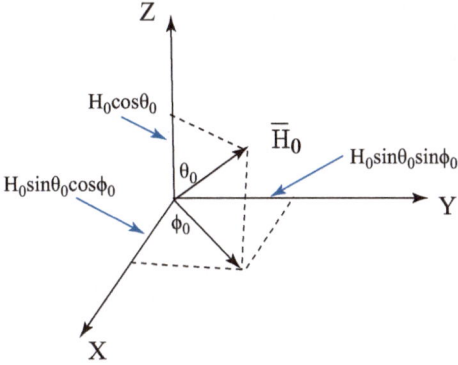

Figure 2.11: Illustration of an arbitrarily oriented DC biasing magnetic field.

Here, the case of $\bar{H}_0$ in the direction (θ_0, φ_0) is taken from Tyras [35]:

$$
[\mu] = \left[
\begin{array}{cc}
\mu + (\mu_0 - \mu)\sin^2\theta_0\cos^2\varphi_0 & \frac{\mu_0-\mu}{2}\sin^2\theta_0\sin 2\varphi_0 + jk\cos\theta_0 \\[4pt]
\frac{\mu_0-\mu}{2}\sin^2\theta_0\sin 2\varphi_0 - jk\cos\theta_0 & \mu + (\mu_0 - \mu)\sin^2\theta_0\sin^2\varphi_0 \\[4pt]
\frac{\mu_0-\mu}{2}\sin 2\theta_0\cos\varphi_0 + jk\sin\theta_0\sin\varphi_0 & \frac{\mu_0-\mu}{2}\sin 2\theta_0\sin\varphi_0 - jk\sin\theta_0\cos\varphi_0
\end{array}
\right.
$$

$$
\left.
\begin{array}{c}
\frac{\mu_0-\mu}{2}\sin 2\theta_0\cos\varphi_0 - jk\sin\theta_0\sin\varphi_0 \\[4pt]
\frac{\mu_0-\mu}{2}\sin 2\theta_0\sin\varphi_0 + jk\sin\theta_0\cos\varphi_0 \\[4pt]
\mu_0 - (\mu_0 - \mu)\sin^2\theta_0
\end{array}
\right] . \tag{2.27}
$$

Note that there is a conjugate symmetry in the off-diagonal elements of the permeability tensor of Eq. (2.27), e.g., $\mu_{ji} = \mu_{ij}^*$ for $i, j = x, y, z$. Also note that when dealing with the permeability tensor, one should carefully handle the time-harmonic dependence, assumed to be either $e^{+j\omega t}$, (which is more common), or $e^{-j\omega t}$, (as some authors do). In Eq. (2.27), a dependence of $e^{+j\omega t}$ is assumed. In those cases, in which the choice is $e^{-j\omega t}$, the sign of jk is reversed.

2.6.6   PERMEABILITY TENSOR: TAKING LOSSES INTO ACCOUNT

Microwave ferrite losses are accounted for in magnetization equation in some of the alternative formal treatments given in Eqs. (2.12), (2.13), and (2.16). It should be noted that a clear understanding of the loss mechanism is difficult to be achieved [12]. Another important observation is that in the above equations, the additional loss term $\bar{M}/\tau$ or $\bar{M} \times (d\bar{M}/dt)$ is perpendicular to $\bar{M} \times \bar{H}$. This represents energy storage, or equivalently, a space lag of 90°. However, since $\bar{M}$ rotates about its equilibrium orientation, this space lag of 90° is equivalent to a time lag of a quarter of the period [12]. This, in turn, yields a complex susceptibility tensor $[\mathcal{X}]$, whose real part represents energy storage and imaginary part denotes losses. Baden Fuller [12] has presented a solution of the magnetization Eq. (2.12). The solution states that for the lossy case, expressions for susceptibility and permeability are similar to those in Eqs. (2.17) and (2.21). In fact, it is provided that resonance frequency ω_0 is substituted with a complex value

$$\omega_0 \leftarrow (\omega_0 + j\omega\alpha), \tag{2.28}$$

where α is the damping factor.

The above is exactly what should be expected from previous explanation about losses causing a quarter-period time lag. Therefore, the permeability tensor is the same for any case given in the previous sections, but μ and k are [12],

$$\mu = \mu_0 + \frac{\gamma\mu_0 M_s \left(\gamma\mu_0 H_0 + j\mu_0\omega\alpha\right)}{\left(\gamma\mu_0 H_0 + j\mu_0\omega\alpha\right)^2 - \omega^2} = \mu_0 \left\{1 + \frac{\omega_m \left(\omega_0 + j\alpha\omega\right)}{\left(\omega_0 + j\alpha\omega\right)^2 - \omega^2}\right\} \tag{2.29}$$

$$k = \frac{\gamma\mu_0 M_s \cdot \omega}{\left(\gamma\mu_0 H_0 + j\mu_0\omega\alpha\right)^2 - \omega^2} = \mu_0 \frac{\omega \cdot \omega_m}{\left(\omega_0 + j\alpha\omega\right)^2 - \omega^2}, \tag{2.30}$$

where ω_0 and ω_m are taken from Eq. (2.19) and the damping factor α is estimated from the measurements of microwave losses around the gyromagnetic resonance ($\omega = \omega_0$). Detailed expressions for the complex susceptibility tensor $[\mathcal{X}]$, when losses are included, are given in [12, 14]. A graphic representation of the real and imaginary part of these complex susceptibilities is shown in Figure 2.12. These curves are obtained by varying either the frequency of the RF microwave signal (ω) or the DC biasing magnetic field H_0, which corresponds to a variation of $\omega_0 = \gamma\mu_0 H_0$. However, it should always be ensured that the ferrite is operated at its saturated state, e.g., $H_0 > 4\pi M_S$.

The gyromagnetic resonance phenomenon occurs when the forced precession frequency is equal to the Larmor free precession frequency ω_0, namely, $\omega = \omega_0$. When losses are not accounted for, μ and k tend to infinity at $\omega = \omega_0$. In contrast, when losses are taken into account, the permeabilities or the susceptibilities become maximum but remain finite at gyromagnetic resonance, as is obvious from Eqs. (2.29) and (2.30), or from Figure 2.5. Note that:

$$j\mathcal{X}_{xy} = j\left(\mathcal{X}'_{xy} - j\mathcal{X}''_{xy}\right) = \mathcal{X}''_{xy} + j\mathcal{X}'_{xy}. \tag{2.31}$$

The damping factor α can be obtained from the imaginary part (representing losses) of the susceptibility near resonance (which is usually measured) and is related to the so-called "resonance linewidth ΔH or $\Delta\omega$."

As shown in Figure 2.13, the resonance linewidth is the width of the resonance curve between the points where the magnitude of $\mathcal{X}''_{xx}$ (or $\mathcal{X}''_{xy}$ or k) becomes half its maximum value. Under the approximation of small losses $\alpha \ll 1$, (common for microwave ferrites), $\alpha^2 + 1 \approx 1$ and the maximum values at resonance $\omega = \omega_0$ are the following [12, 14]:

$$\text{for} \quad \omega = \omega_0 \leftrightarrow \mathcal{X}''_{xx,\text{max}} = k''_{\text{max}} = \frac{\omega_m}{2\alpha \cdot \omega_0}. \tag{2.32}$$

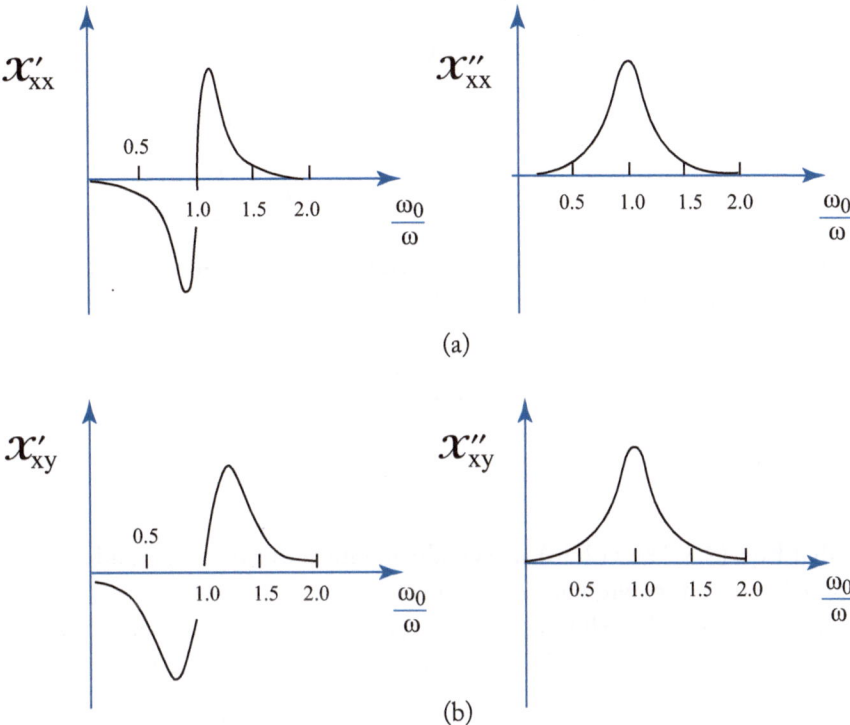

(a)

(b)

Figure 2.12: Graphic representation of typical ferrite complex susceptibilities.

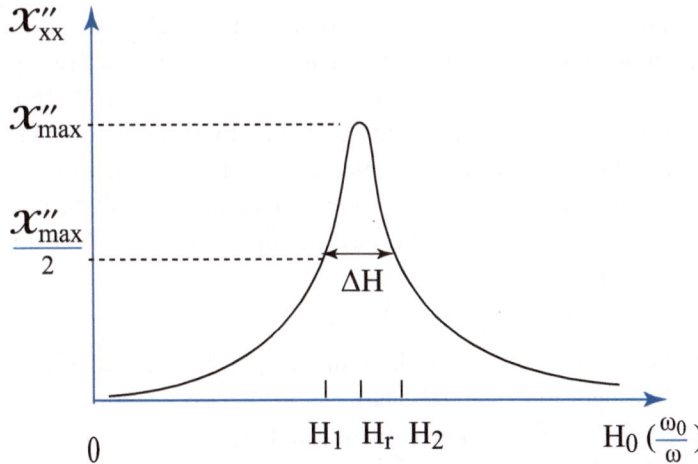

Figure 2.13: Gyromagnetic resonance and definition of the resonance linewidth (ΔH or $\Delta \omega_0$).

Furthermore, the resonance linewidth is [12]:

$$\Delta\omega = \gamma\mu_0\Delta H = 2\omega_0\alpha. \tag{2.33}$$

Since ΔH can be measured, the damping factor can be estimated by, e.g., [37]:

$$\alpha = \frac{\gamma\mu_0\Delta H}{2\omega_0} \quad \text{or} \quad \omega\alpha = \frac{1}{2}\frac{\omega}{\omega_0}\gamma\mu_0\Delta H = \frac{1}{2}\omega\frac{\Delta H}{H_0}. \tag{2.34}$$

However, when microwave frequency is assumed to be constant and the DC magnetic field is changed by ΔH, according to Pozar [14], it is:

$$\Delta H = \frac{\Delta\omega_0}{\mu_0\gamma} = \frac{2\alpha\omega}{\mu_0\gamma}. \tag{2.35}$$

So,

$$\alpha\omega = \mu_0\gamma\frac{\Delta H}{2}. \tag{2.36}$$

Note that Eqs. (2.35) and (2.36) are actually identical to Eq. (2.33), and bear in mind that ΔH is measured around the gyromagnetic resonance $\omega = \omega_0$.

Typical prinel ferrites have linewidth ΔH ranging from 100–500 oersted, while for ferrite garnets, like YIG, $\Delta H < 100$ oersted. Single YIG crystal presents linewidth as low as 0.3 oersted.

2.7 DIELECTRIC PROPERTIES OF FERRITES

The dielectric constant (ε_r) of ferrites is usually assumed to be constant and any frequency dispersion is usually omitted. It will be shown that this is a reasonable hypothesis at least within the frequency and temperature limits in which ferrites are used in practice. Spinel ferrites have a typical ε_r from 12–13, while in ferrimagnetic garnets ε_r range from 14–16 [38].

Electric polarization in ferrites is due to the same physical phenomena as those of other solids. Therefore, we will briefly review these mechanisms here by focusing on ferrites' specific composition. When a material is subject to an external electric field, there are basically four mechanisms producing electric dipoles. The electric dipole moments per unit volume define electric polarization $\bar{P}$ [14]:

$$\bar{P} = N \cdot \alpha_T \cdot \bar{E}_i, \tag{2.37}$$

where N is the number of molecules per unit volume, $\bar{E}_i$ is the internal electric field, and α_T is the average polarizability per molecule. The latter is composed of four terms corresponding to the contribution of the four different mechanisms, namely,

- *electronic polarizability (α_e)*

- *ionic polarizability (α_i)*

- *permanent dipole polarizability* (α_d)

- *space charge polarizability* (α_s)

$$\alpha_T = \alpha_e + \alpha_i + \alpha_d + \alpha_s. \tag{2.38}$$

A typical graph of the relative contributions of different polarizabilities vs. frequency is shown in Figure 2.14. Each one of them will be briefly reviewed below.

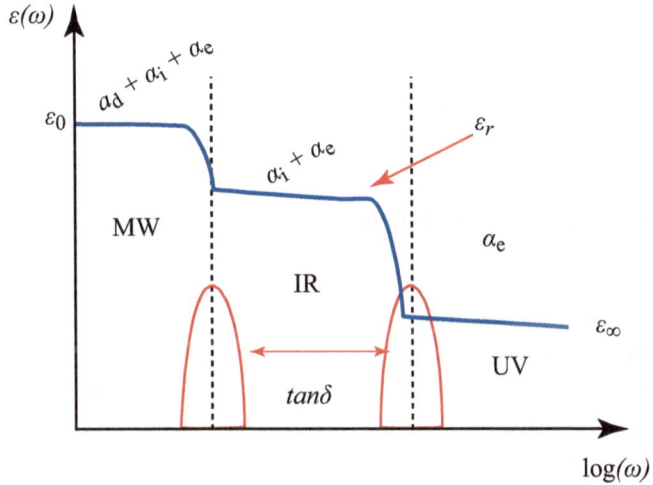

Figure 2.14: Relative contributions of the polarizability mechanisms vs. frequency.

2.7.1   ELECTRONIC POLARIZABILITY (α_e)

The application of an electric field causes an elastic displacement of electrons with respect to the nucleus of each atom, distorting electron shells as shown in Figure 2.15. The induced dipole moments give rise to electronic polarization. As shown in Figure 2.14, only this mechanism is effective from zero frequency up to the optical spectrum and undergoes relaxation in the ultraviolet region.

The contribution of electronic polarizability to the dielectric constant of ferrites is significant since, in the optical region a value of ε_r from 5–7 is typically observed. Most of this contribution is due to oxygen atom polarizability, whose α_e is much larger than that of metallic atoms.

2.7.2   IONIC OR ATOMIC POLARIZABILITY (α_i)

The applied electric field also acts on anions as well as cations, and it may cause an elastic separation within the molecule as sketched in Figure 2.15. The above mechanism produces effective

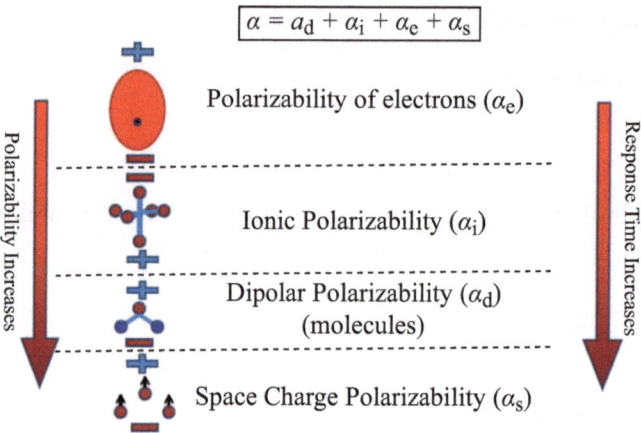

Figure 2.15: Physical mechanisms producing electric dipoles within a medium for electronic, ionic, permanent dipoles, and space charge polarizabilities [38].

dipoles, which give rise to ionic or atomic polarization (α_i). This effect contributes to the total polarization up to the infrared region, where it undergoes relaxation, as shown by α_i in Figure 2.14. In ferrites and garnets, ionic polarizability increases the dielectric constant to ($\alpha_e + \alpha_i$), which in the range of microwave frequencies is from about 12–16 [37].

2.7.3   PERMANENT DIPOLE POLARIZILITY (α_d)

The constituent ions of many molecules have different affinities with electrons. The asymmetry caused by these different charges gives rise to permanent dipole moments, and these molecules are called "polar molecules." In the absence of an electric field, these permanent dipoles are randomly oriented, and their effect is canceled out, as shown in Figure 2.14. When an electric field is applied, these dipoles tend to line up with it and the degree of alignment depends on the type of substance used (e.g., they are completely aligned in fluids). This is permanent dipole polarizability (α_d), which, as shown in Figure 2.14, is effective below the far infrared frequencies. The contribution of this mechanism in the dielectric constant of spinel ferrites and garnets is usually negligible.

2.7.4   SPACE CHARGE POLARIZABILITY (α_s)

In a lot of materials, including ferrites, there are charge carriers (e.g., conduction electrons) free to migrate through the medium. However, their motion may be blocked by imperfections or boundary surfaces creating space charges, which result in field distortion, as shown in Figure 2.15. The application of an electric field gives rise to a polarizability (α_s) term, which is effective only in the low frequency range (far below the RF and microwave bands). Very large dielectric constants of the order of 10^5 are attributed to this mechanism. This is also effective in ferrites and garnets, but only at very low frequencies [38].

In summary, the dielectric constant of ferrites and garnets operating in the RF microwave region is mainly due to the electronic (α_e) and ionic (α_i) polarizabilities.

2.7.5   DIELECTRIC LOSSES

All the above mechanisms contributing to the real part of the dielectric constant are accompanied by frictional loss phenomena. They are associated with the movement of charges and relaxation mechanisms when the electric field alternates. In order to account for these losses, the relative contributing factors (α_e), (α_i), (α_d), and (α_s) become complex, and so does the dielectric constant.

There are losses due to the conductivity of the medium, which are also accounted for by the imaginary part of the complex dielectric constant $\varepsilon^* = \varepsilon' - j\varepsilon' = \varepsilon_r - j\varepsilon_r \tan \delta_\varepsilon$. For most ferrites and garnets, the AC resistivity (ρ) is greater than 10^6 ohm-cm, and the expected loss tangent ($\tan \delta_\varepsilon$) in the microwave range is less than 2×10^{-4} for garnets and from 5×10^{-4} to 2×10^{-3} for spinel ferrites [38].

It should also be noted that, although the above mechanisms are in general non-linear, for most spinel ferrites and garnets, these effects are linear for fields up to the breakdown level of the material [14].

2.7.6   CONDUCTION MECHANISM IN FERRITES AND GARNETS

It is interesting to examine the conduction mechanism in ferrites and garnets, which is mainly due to the migration of electrons through the crystals [11]. This is, in turn, explained by the change of iron cations from a divalent to a trivalent state; in other words, at any dislocation in the crystal structure, divalent iron cations may easily change to a trivalent state and vice-versa. This involves an electron migration that gives rise to a conduction mechanism and, thus, conduction losses. Therefore, iron lean ferrites present losses due to higher resistivity and, thus, lower conductivity. This happens at the expense of lower saturation magnetization. In order to preserve high saturation magnetization, an iron rich ferrite is required. Therefore, *the ideal ferrite or magnetic garnet should have only trivalent iron ions* if we want to minimize conduction losses. This also reduces magnetic losses as will be explained next.

Representative examples involving only trivalent iron ions are lithium ferrite ($(LiFe)_{1/2}Fe_2O_4$) and YIG ($Y_3Fe_5O_{12}$), which indeed have the lowest microwave losses. It should also be noted that in ferrites the number of iron ions should be less than the stoichiometric ratio in order for them to be in a trivalent state; otherwise, some divalent ions may occur.

Moreover, the production of low-loss ferrites should ensure adequate purity and appropriate temperature conditions. Impurities can have detrimental effects on resistivity [12]. Due to impurities, a monovalent ion may occupy a divalent atom position, causing a semiconductor effect. Likewise, divalent ions may occupy trivalent atom positions, leaving floating electrons or supervalent ions. In any of these cases, conductivity and the related losses are increased.

2.7.7   MAGNETIC LOSSES

Concerning magnetic losses, the corresponding damping factor is strongly affected by spin-orbit coupling. According to [12], trivalent iron ions have the lowest damping due to the absence of spin-orbit coupling since their magnetic moment is due to five electron spins occupying the third shell and lack of orbital moment.

2.8   FERROELCTRIC PROPERTIES

Ferroelectrics are in many respects similar to ferrites. To be more specific, they are dual of ferrites since they exhibit tunable permittivity instead of tunable permeability. A major difference between them pertains to anisotropy; while in ferrites, magnetic anisotropy constitutes a major effect, electric anisotropy does not usually play an essential role in the performance of tunable ferroelectrics [39]. However, the elementary electric dipole moments, which give rise to polarizability in ferroelectrics, form domains (small regions) where dipoles are completely aligned in a certain direction just like in ferrites. Moreover, the application of an external electric field causes phenomena similar to those in ferrites. According to [39], ferroelectrics are known for their high dielectric constant and high tunability. In general, tunability is higher for higher dielectric constants. The above holds at least for an "ideal ferroelectric," whose dielectric response is controlled by the material lattice dynamics.

2.8.1   ELECTRIC POLARIZATION–PERMITTIVITY

The physical mechanisms giving rise to electric polarization are in general the same for any material. They have already been described above in connection with the dielectric properties of ferrites. Thus, electric polarization is composed of "electronic," "ionic," "permanent dipole or orientational," and "space charge" polarization. As explained in [40, 41], *electronic polarization* is an induced effect due to the application of an electric field. This field causes a displacement of the center of gravity of negative and positive charges within an atom (see Figure 2.16a), giving rise to induced dipole moments. Likewise, the application of an electric field causes elastic deformations among the bonds of anions and cations, producing effective dipoles (Figure 2.16b). These dipoles give rise to "*ionic or atomic polarization.*" As shown in Figure 2.16b, depending on the direction of the applied field, anions and cations move either closer to each other or further apart [41]. Besides ionic polarization, a change of the material's overall dimensions may occur; namely, "piezoelectric effects" may accompany polarization. *Space charge polarization* is observed in materials in which there are charge carries free to migrate through the medium. This phenomenon is insignificant in ferroelectrics and is usually omitted.

Orientational polarization is the key phenomenon contributing to ferroelectric properties, including tunability. Recall that in many molecules, their constituent ions have different affinities of electrons, and this asymmetry in charge gives rise to permanent dipoles. Well-known examples are water and some oils used in transformers, where one end of the molecules is effectively positive and the other end is effectively negative [41]. Thermal agitation tends to randomize these dipole orientations. The application of an electric field tends to align them along the field's direction as shown in Figure 2.16c [41]. However, orientational polarization in ferroelectrics is due to asymmetrically located ions in the crystal structure.

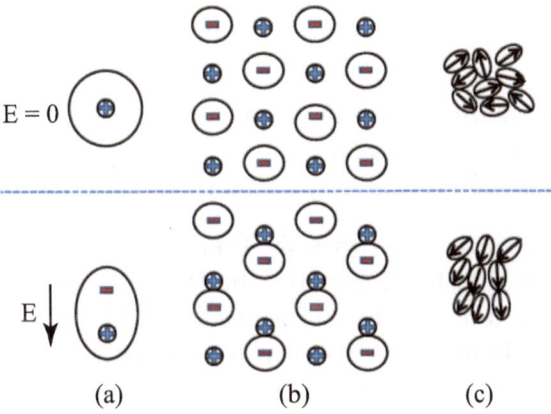

Figure 2.16: Schematic representation of the physical mechanisms producing electric dipoles for: (a) electronic, (b) ionic or atomic, and (c) orientational polarizabilities.

The relative contribution of the three polarization mechanisms to the dielectric constant is presented in Figure 2.17. It can be observed that orientational polarizability is effective *below the far infrared*. Ionic polarizability contributes up to the infrared region, and electronic polarizability is observed in almost the whole frequency spectrum—specifically up to the ultraviolet region. This dependence of the dielectric constant is also known as the material frequency dispersion. The loss tangent is also presented at some indicative frequencies in Figure 2.17. It is obvious that each polarization mechanism is accompanied by corresponding losses. These losses sum up at the lower part of the spectrum, where all three mechanisms are effective. Alternatively, it can be said that a higher dielectric constant is accompanied by higher losses.

2.9   FERROELECTRICITY

Electronic and ionic polarizations revert to an unpolarized state when the electric field is removed. Materials supporting only electronic and ionic polarizations are called "***nonpolar materials.***" This is because they do not exhibit individual electric dipoles in the absence of an applied electric field. In contrast, orientational polarization can be divided into two classes. One class

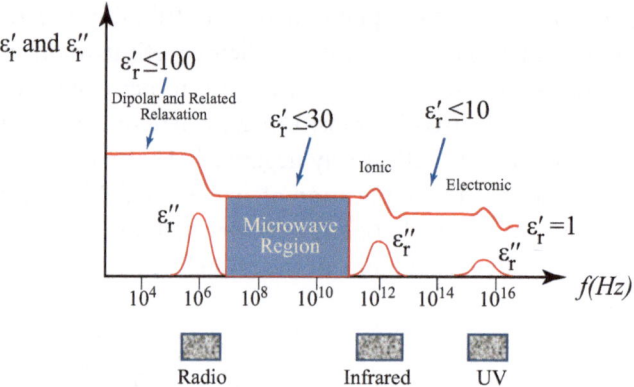

Figure 2.17: The relative contribution of the three polarizability mechanisms to the real and imaginary part of the dielectric constant (ε_r).

can be reverted to an unpolarized state when the field is removed, and the other, the *remnant class*, retains polarization after the field is removed. The latter is usually reversible; in other words, the polarization direction is reversed when the applied field changes sign. In general, materials exhibiting a non-zero net electric dipole moment and a corresponding polarization in the absence of an applied electric field are called "*electrets*." On the other hand, materials exhibiting an individual electric dipole, which in the absence of an applied field are arbitrarily oriented to yield a zero net total dipole moment and zero polarization, are called "*polar materials*."

Finally, materials that retain a reversible net polarization when the field is removed are called "*ferroelectrics*" [40, 41]. In a manner similar to that of ferrites, ferroelectrics exhibit small regions with strongly coupled dipole domain moments, also called "*domains*." This strong coupling causes the dipole moments within a domain to be aligned in the same direction. Once again, as in ferrites, the natural tendency of stored energy minimization leads to multiple domains which have different orientations so that a freshly annealed ferroelectric does not exhibit spontaneous polarization. When an electric field is applied, the domain walls are moved in such a way as to enlarge the domains, whose polarization components are in the direction of the applied field. A further increase in the applied electric field tends to rotate the remaining domain dipole moment as a whole in the direction of the applied field. In a manner similar to that in ferrites, as the electric field is increased, some domains are easily aligned. However, fewer additional domains are aligned as the field is further increased, which causes a saturation effect.

Saturation polarization depends on temperature, and above a certain characteristic temperature, called the "*Curie-Weiss temperature*" (T_C), the thermal vibration-agitation of atoms becomes a dominant phenomenon and saturation polarization tends to zero; that is, the material loses its ferroelectric properties. This behavior is described as "*paraelectric*."

2.10   HYSTERESIS LOOP

The application of an alternating electric field to a ferroelectric material shows that the polarization traces a hysteresis loop as shown in Figure 2.18. Recall that when an increasing DC electric field is applied to an unpolarized ferroelectric, the energetically favorable domains (those with dipoles aligned in the applied field direction) start expanding with the appropriate movement of the domain walls. In this case, polarization follows the curve OB in Figure 2.18. At point B and beyond, the whole ferroelectric specimen becomes an almost monodomain crystal where all dipoles are aligned along the applied field.

A further increase in the field beyond point B and up to a maximum intensity Em, corresponding to the maximum voltage applied to the crystal, causes a linear increase in polarization along curve BC, which is due only to the mechanisms of induced dipoles (electronic and ionic). Decreasing the field from Em toward zero causes polarization to follow a reversible part CB first (the one due solely to induced dipoles), and then the irreversible trace BD. At zero-applied electric field, point D, the crystal remains monodomain with a remnant polarization P_S [40]. Changing the sign of the applied field and increasing its intensity toward negative values causes polarization to follow trace DF; that is, polarization remains positive and decreases slowly from its remnant value. At some point F, the electric field reaches some critical value (-Ec), called "the coercive field." At this point, the regions (domains) with spontaneous polarization parallel to the (negative) field direction start to expand rapidly. This causes a rapid decrease in polarization, a change of the sign, and an "increase" toward negative values following curve FL. At point L, the crystal becomes monodomain again and polarization is aligned along the applied electric field. Similar phenomena occur when the electric field starts to increase either toward negative or positive values. Finally, an alternating electric field will cause the polarization vector to follow the hysteresis loop of Figure 2.18. As usual, the area enclosed within the hysteresis loop represents a loss of energy, known as hysteresis loss. However, the application of a strong DC electric field in conjunction with a weak RF microwave alternating field will result in the small area of the hysteresis loop shown in Figure 2.19.

2.11   FERROELECTRIC MATERIALS—PEROVSKITES

The phenomenon of ferroelectricity was discovered by Valasek in 1920 [42] while he was investigating the anomalous dielectric properties of the Rochelle salt ($NaKC_4O \cdot 6\,H_2O$). Many different ferroelectric materials have been discovered since that early work. Most of the practically used ferroelectrics exhibit a so-called "perovskite" crystal structure, which is a corner-sharing octahedron, presented in Figure 2.20. It may seem strange, but the perovskite crystal structure is also encountered in some ferrite types [43].

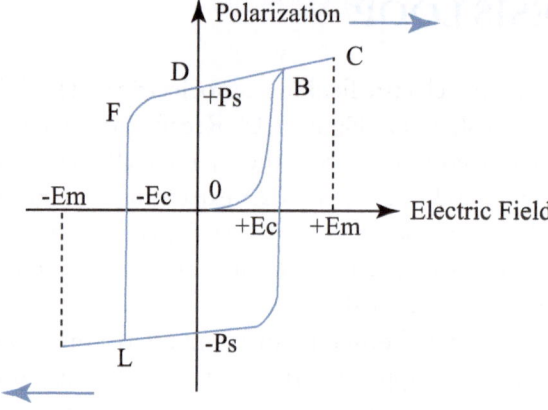

Figure 2.18: Polarization curves of a ferroelectric crystal and its hysteresis loop when a periodic electric field is applied.

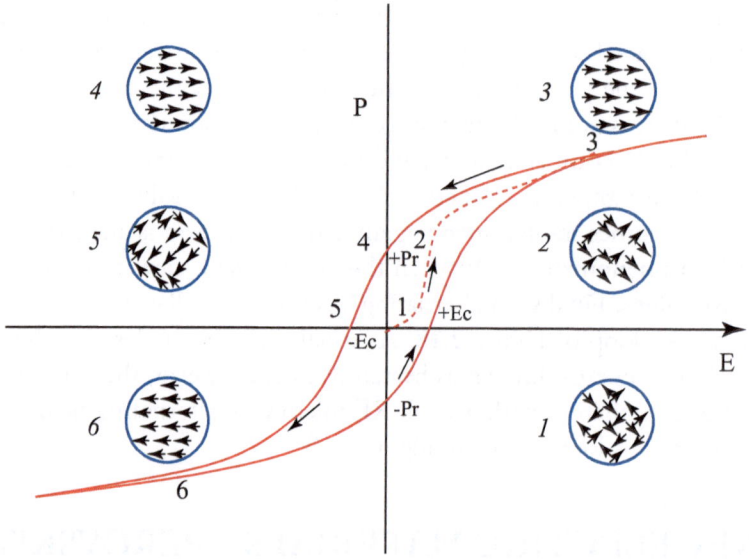

Figure 2.19: Ferroelectric hysteresis loop under small periodic electric field conditions.

2.12 THE PEROVSKITE CRYSTAL STRUCTURE

The general formula of the perovskite crystal structure is ABX_3, where A and B are cations and X is an anion, usually oxygen. For ferroelectrics, it is $A = (Sr, Ba, Y, Ca, \ldots)$, $B = (Ti, Cu, Ru, \ldots)$ and $X = (O, F, \ldots)$. As shown in Figure 2.20, the A sites are 12-fold co-ordinated by X atoms (usually $X = O$ oxygen), the B sites are 6-fold co-ordinated, and each X

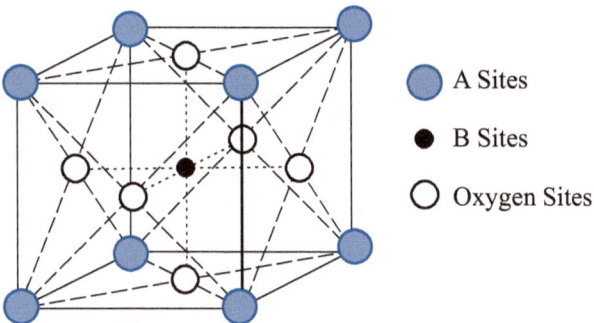

Figure 2.20: Unit cell of the perovskite crystal structure [40, 52, 53].

anion is coordinated by four (4) A sites and two (2) B sites. According to [40–48], the geometrical requirements for a perovskite crystal to be realized were first studied by Goldschmidt [44] as:

$$R_A + R_X = t\sqrt{2}\,(R_B + R_X),\qquad(2.39)$$

where R_A, R_B, and R_X are the radii of ions A, B, and X, respectively.

In an ideal perovskite structure $t = 1$. However, in nature, the radii do not exactly obey Eq. (2.39) and a certain amount of distortion is possible. Goldschmidt [44] has shown that the range of t is $0.85 \leq t \leq 1.05$ and the ratio of the ionic radii can vary between $0.41 < R_B/R_X < 0.73$ and $R_A/R_X < 0.73$.

A widely known ferroelectric is $SrTiO_3$, where $A = Sr^{+2}$, $B = Ti^{+4}$, and $X = O^{-2}$. Its perovskite crystal is presented in Figure 2.21a [40]. The oxygen atoms occupy the positions on the facets of the cube, more specifically the middle of each edge. Strontium (Sr^{+2}) atoms occupy the cube's center positions and titanium (Ti^{+4}) atoms reside at its corners. One of the most important ferroelectrics, widely used and extensively studied for microwave applications, is barium strontium titanate (BSTO); ($Ba_X\,Sr_{1-X}\,TiO_3$), which also exhibits a perovskite crystalline structure. This crystal is derived from that of $SrTiO_3$ by replacing a fraction (x) of Sr^{+2} atoms with Ba^{+2} atoms. This is practically obtained as a solid solution of $BaTiO_3$ and $SrTiO_3$. An important property is that its Curie temperature T_C, namely the critical temperature at which the transition from a ferroelectric to a paraelectric state occurs, is controlled by the quantity-percentage of Ba^{+2} atoms. That is, T_C can be tuned by Ba doping.

2.13 FERROELECTRICITY AS A RESULT OF CRYSTALLIC ASYMMETRY

Ferroelectricity stems from the absence of a center of symmetry in the perovskite crystal structure. This asymmetry produces an effective electric dipole that gives rise to spontaneous orientation polarization [48–55]. As shown in Figure 2.22a, which depicts the crystal structure of

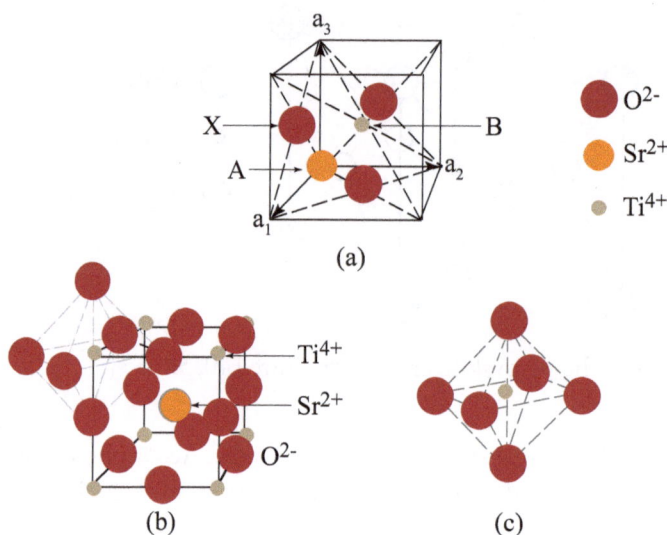

Figure 2.21: Perovskite crystalline structure. (a) A and B are cations and X is anion, (b) $SrTiO_3$, and (c) oxygen octahedron.

Barium titanate ($BaTiO_3$), the Ti^{+4} ion, is slightly displaced (≈ 0.12 Å) from its ideal centered position. The Ba^{+2} and O^{-2} ions are also displaced, and the resulting crystal structure is classi-fied as tetragonal symmetry. However, when the temperature is increased above a certain critical value (the Curie–Weiss temperature T_C), the thermal agitation of atoms yields a highly symmet-ric crystal lattice (cubic symmetry), as shown in Figure 2.22b. Due to this high symmetry, the crystal does not present any orientational electric dipole moment and, as a result, spontaneous polarization disappears.

Below the Curie–Weiss temperature, the material is said to be in its ferroelectric phase since its spontaneous polarization leads to hysteretic behavior exhibiting permanent polariza-tion. In contrast, above the Curie temperature, the material is paraelectric with spontaneous polarization and its effects disappear. *However, it is important to note that the dielectric constant dependence on the applied electric field DC bias and temperature is retained in the paraelectric phase;* that is, the material is still tunable. In fact, it is mostly in this paraelectric phase that these materials are employed in microwave applications, exploiting the advantage of the absence of hysteresis losses.

2.14   PARAELECTRIC PHASE

When the temperature of a ferroelectric reaches its critical Curie–Weiss value (T_C), the ma-terial undergoes a transition from the ferroelectric to the paraelectric phase, where permanent polarization disappears. In the ferroelectric phase, the crystal lattice is tetragonal and the electric

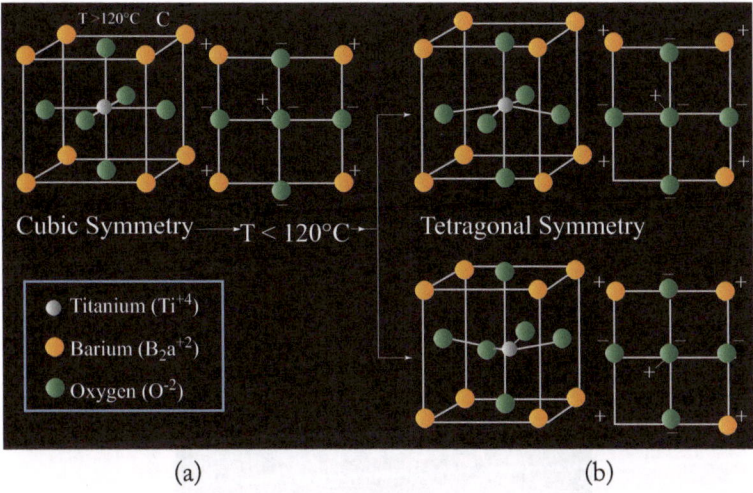

Figure 2.22: The crystallic perovskite structure of $BaTiO_3$: (a) Cubic lattice symmetry of a highly symmetric perovskite structure above the Curie–Weiss temperature $> T_C$. (b) Tetragonal lattice symmetry of a distorted (asymmetric) perovskite structure below the Curie–Weiss temperature $T < T_C$.

polarization response to an alternating electric field follows a hysteresis loop as shown in Figure 2.23. When the temperature rises above the Curie–Weiss point ($T = T_C$), the crystal lattice changes from tetragonal (Figure 2.23a) to cubic (Figure 2.23b) and polarization is not hysteretic (Figure 2.23c) anymore, as shown in Figure 2.23d. This is a highly desirable property since the operation in the paraelectric state will be free of hysteresis losses.

Another phenomenon of equal importance is the temperature dependence of the dielectric constant around the Curie–Weiss point. Figure 2.24 shows the dielectric constant vs. temperature in the absence of a DC biasing electric field $E(0)$. It increases up to the Curie point, where it reaches its maximum value. Then it starts to decrease very fast at first, and then slowly [46]. This temperature dependence remains the same when a DC biasing electric field is applied, but the maximum of the dielectric constant is reduced with an increase in the DC biasing field.

It is important to observe in Figure 2.24 that the temperature dependence of thick films (with thickness greater than about 50 nm) is similar to that of the bulk material. However, this behavior is almost suppressed in thin films, which seem to remain in the paraelectric phase and whose dielectric constant is greatly reduced. For example, BSTO thin films have a dielectric constant ranging from $\varepsilon_r = 200$ to 300 [46].

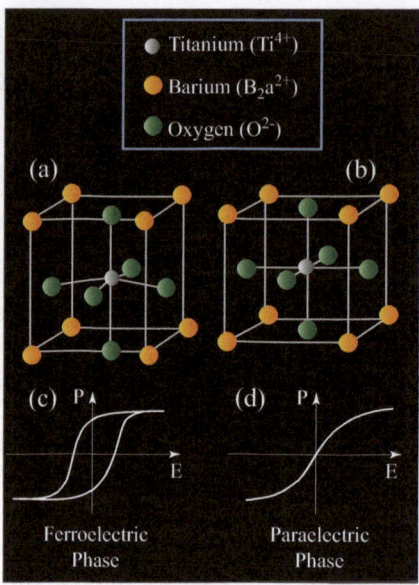

Figure 2.23: Crystal and electric polarization response in the ferroelectric and paraelectric states: (a) "tetragonal" lattice below T_C, (b) "cubic" lattice above T_C, (c) hysteresis loop in the ferro-electric phase, and (d) non-hysteretic polarization in the paraelectric phase [46].

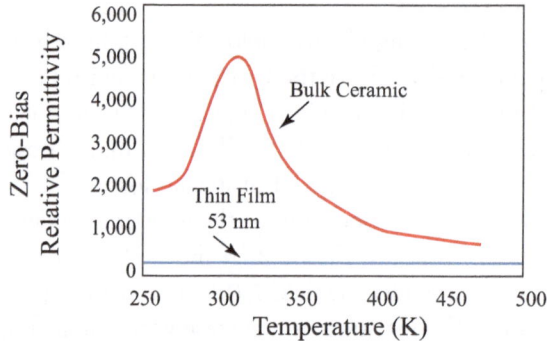

Figure 2.24: Temperature dependence of the dielectric constant for bulk or thick-film and thin-film BSTO [42, 44, 46].

2.15 QUANTUM OR INCIPIENT FERROELECTRIC

Quantum or incipient ferroelectrics are those with a Curie–Weiss temperature T_C tending to absolute zero. However, the temperature of the impending phase transition (lower than about 4 K), from paraelectric to ferroelectric, may not be realized. This is because the temperature is low enough to activate quantum effects [48]. For this reason, they are called "quantum ferroelectrics."

Typical representatives of quantum ferroelectrics are the perovskite crystals of STO ($SrTiO_3$), $CaTiO_3$, and KTO ($KTaO_3$). When an amount (x) of ppm, (parts per million), of impurities is added to quantum ferroelectrics, the Curie–Weiss temperature (T_C) is increased even up to room temperature or higher. BSTO ($Ba_xSr_1-xTiO_3$) below T_C exhibits a typical-conventional ferroelectric-like behavior [48]. For this reason, quantum ferroelectrics are also called "incipient ferroelectrics" since they constitute the basic component of which conventional ferroelectrics are composed. The term "incipient ferroelectric" means that the material manifests dielectric non-linearity but does not exhibit spontaneous polarization [48, 49]; that is, it has no polar phase. According to Ang et al. [49], and the references therein, quantum paraelectrics show an increase in the dielectric constant up to about 4 K, and then exhibit a levelling-off around 0 K. A typical temperature dependence of the dielectric constant of incipient dielectrics for different DC biasing electric fields is shown in Figure 2.25 [48, 49, 51].

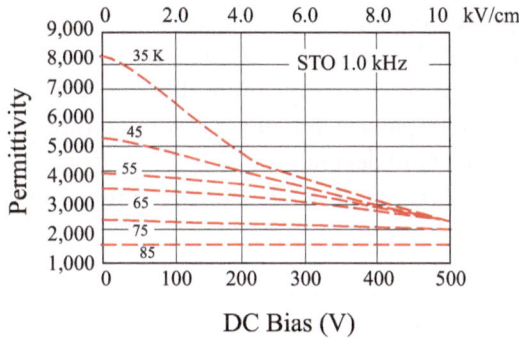

Figure 2.25: Temperature dependence of the dielectric constant for the incipient dielectric $SrTiO_3$, (STO), vs. the DC bias voltage.

2.16   PEROVSKITE SUPERLATTICES

The undesired temperature dependence and the high microwave losses of perovskites in their paraelectric phase have led material scientists to search for more advanced tunable materials. Recently, e.g., [55], perovskite superlattices have been proposed for this purpose. A superlattice comprises alternating, (usually epitaxial), layers of dissimilar paraelectric perovskites and may exhibit large changes (tunability) in the dielectric constant with a variation of the applied DC bias electric field. Not all the layers of a superlattice need to be tunable, but at least one should. Experiments with superlattices composed of any suitable sequence of paraelectric $SrTiO_3$, $SrCeO_3$, $SrZrO_3$, $BaTiO_3$, $BaZrO_3$, $CaZrO_3$, and $LaAlO_3$ layers have been conducted in the literature [56–65]. They were fabricated with the aid of pulsed laser deposition, mostly on a $LaAlO_3$ substrate and some of them on a $SrTiO_3$ one. Superlattices can be made to have weak temperature dependence and large tunability. For example, the $SrTiO_3/BaZrO_3$ superlattice exhibits a tunability of 33%, both at room temperature and at 77 K. In contrast, typical ferroelectrics are tunable only in a narrow temperature range, around the phase transition at the Curie–Weiss temperature. Moreover, while the dielectric constants (ε_r) in perovskites are decreased with an increase in the DC biasing electric field E_{DC}, (namely, negative $d\varepsilon_r/dE_{DC}$), some superlattices exhibit an increasing dielectric constant $d\varepsilon_r/dE_{DC} > 0$ [54, 55].

2.17   CONVENTIONAL FERROELECTRICS—TEMPERATURE AND DC BIAS DEPENDENCE

The dielectric constant of conventional ferroelectrics, e.g., the solid solution BSTO ($Ba_XSr_{1-X}TiO_3$), behaves similarly to that of incipient perovskites in the paraelectric phase above the Curie-Weiss temperature. This is so in terms of both temperature and DC biasing electric field. In the ferroelectric phase, the dielectric constant increases to reach its maximum value at the Curie–Weiss point. Moreover, concerning the BSTO solid solution, the barium percentage (x) controls the Curie–Weiss temperature. For $x < 0.5$, the T_C is below room temperature [51]; thus, BSTO is in its paraelectric phase at room temperature. The typical temperature and the DC biasing electric field dependence of the dielectric constant of BSTO have been given in [57] and can be found in [49, 52]. The above is presented in Figures 2.26 and 2.27, respectively.

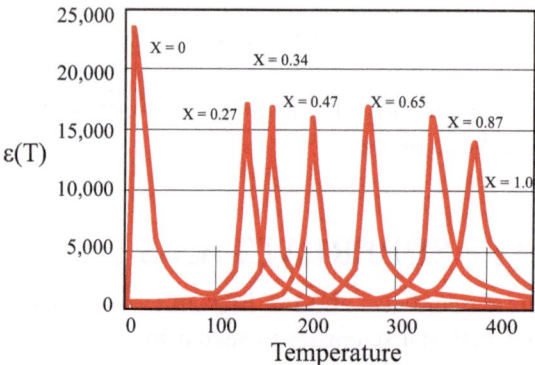

Figure 2.26: The dielectric constant of BSTO, ($Ba_XSr_{1-X}TiO_3$), vs. temperature for different barium contents (x).

2.18   SUPERCONDUCTOR PEROVSKITES

An interesting orthorhombic perovskite-like crystal structure is that of the high-temperature superconductor (HTS) of yttrium barium copper oxygen, (YBCO), ($YBa_3Cu_3O_{7-\delta}$) [40]. Often,

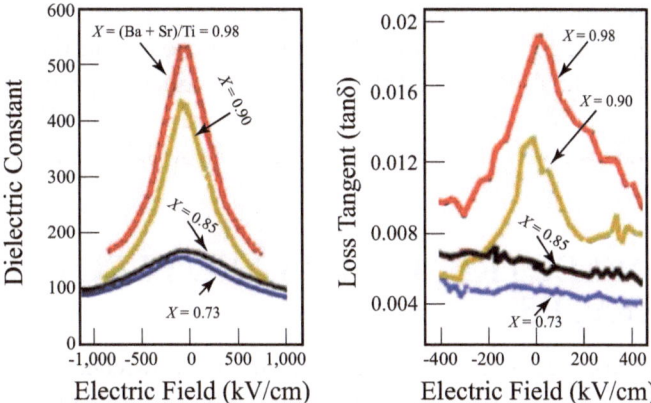

Figure 2.27: Dielectric constant of $Ba_XSr_{1-X}TiO_3$ (BSTO) for $X = (Ba + Sr)/Ti = 0.98, 0.9, 0.85$, and 0.73 vs. the applied DC biasing electric field [49, 52].

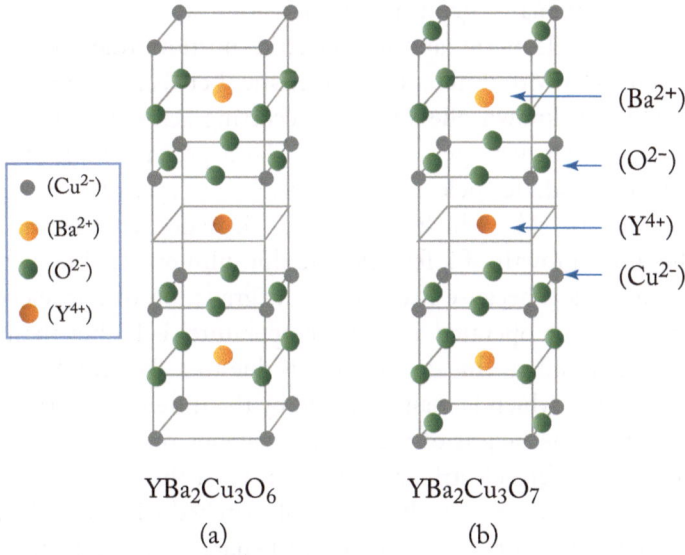

Figure 2.28: An orthorhombic perovskite-like crystal structure of a YBCO superconductor: (a) oxygen-deficient $YBa_2Cu_3O_6$ and (b) fully oxygenated $YBa_2Cu_3O_7$.

YBCO is used to grow thin film electrodes on the top of ferroelectrics. As shown in Figure 2.28, its crystalline structure is composed of three perovskite unit cells stacked upon each other, namely, alternating Ba-Cu-O and Y-Cu-O perovskite unit cells. The oxygen content controlled by a parameter δ strongly affects the conductivity of the material. An oxygen-deficient structure with $\delta = 1$, $YBa_2Cu_3O_6$, and a fully oxygenated crystal structure with $\delta = 0$, $YBa_2Cu_3O_7$, are shown in Figures 2.28a and b, respectively. The latter, (Figure 2.28b, $\delta = 0$), presents the highest conductivity, while for $\delta = 0.6$ to 0.8, a transition occurs and the material ceases to be a superconductor. It should also be noted that the superconducting properties in YBCO are in general anisotropic. In addition, the critical transition temperature of YBCO from a normal to a superconducting state is 93 K. This temperature can be provided by liquid nitrogen that has a boiling point at 77 K.

2.19   FERROELECTRIC LAYERS AND ELECTRODE INTERFACES

In practical applications, a ferroelectric material is either sandwiched between two electrodes to form a tunable capacitor or grown (e.g., as a thin epitaxial film) on the top of a grounded substrate (with a grounded electrode on its bottom side) in order to realize the printed microwave circuit (see Figure 2.29). In the latter case, the electrode may be manufactured, often using electron beam evaporation techniques, either on top of a superstrate or directly on top of the ferroelectric. Phenomena of great importance occur at the following interfaces: ferroelectric/electrode, substrate/electrode, and ferroelectric/substrate (dielectric or ferrite garnet). The formation of a thin layer with different characteristics between a metallic electrode and the dielectric substrate is a well-known phenomenon occurring across the whole frequency spectrum. At the very low frequency range, this parasitic layer presents high impedivity, (high resistivity) [44]. At the electrode/ferroelectric interface, a thin layer of a low dielectric constant is formed at the ferroelectric side, which substantially degrades ferroelectric permittivity. According to [40], the direct growth of a ferroelectric thin film on a noble metal electrode (Au, Pt) should be avoided. This is because it causes the degradation of ferroelectric features even though its conductivity (when operated at room temperature) is higher than that of the preferential metallic oxide electrodes. This degradation is due to a **Schottky barrier** built up at the metal-ferroelectric interface, which is mainly caused by the difference in the work functions of the two media. Moreover, the height of the barrier depends on electron affinity, the density of the charge carriers in the ferroelectric, and the density of the states at the interface. Well-known metallic oxide electrodes, compatible and well matched with ferroelectrics like $SrTiO_3$ or BSTO, are the superconductor YBCO or the ruthenate superconductors SRO ($SrRuO_3$), LSMO ($La_{2/3} Sr_{1/3} MnO_3$), and LCMO ($La_{2/3} Ca_{1/3} MnO_3$). However, the high resistivity of these metallic oxide electrodes yields high microwave losses. A solution to this problem is the use of the metallic oxides as buffer-matching layers between ferroelectrics and noble metal electrodes as shown in Figure 2.29. A very promising structure offering good film quality, high degree of tunability, and low microwave losses is Au/Pt/BSTO/SRO/Pt/Au [40].

A block diagram of the low-temperature co-fired ceramic (LTCC) technology is analytically presented in Figure 2.30.

The goal set for the epitaxial growth of the above multilayer thin films is to achieve properties as good as those of single crystal bulk materials do. To this purpose, the layers with a

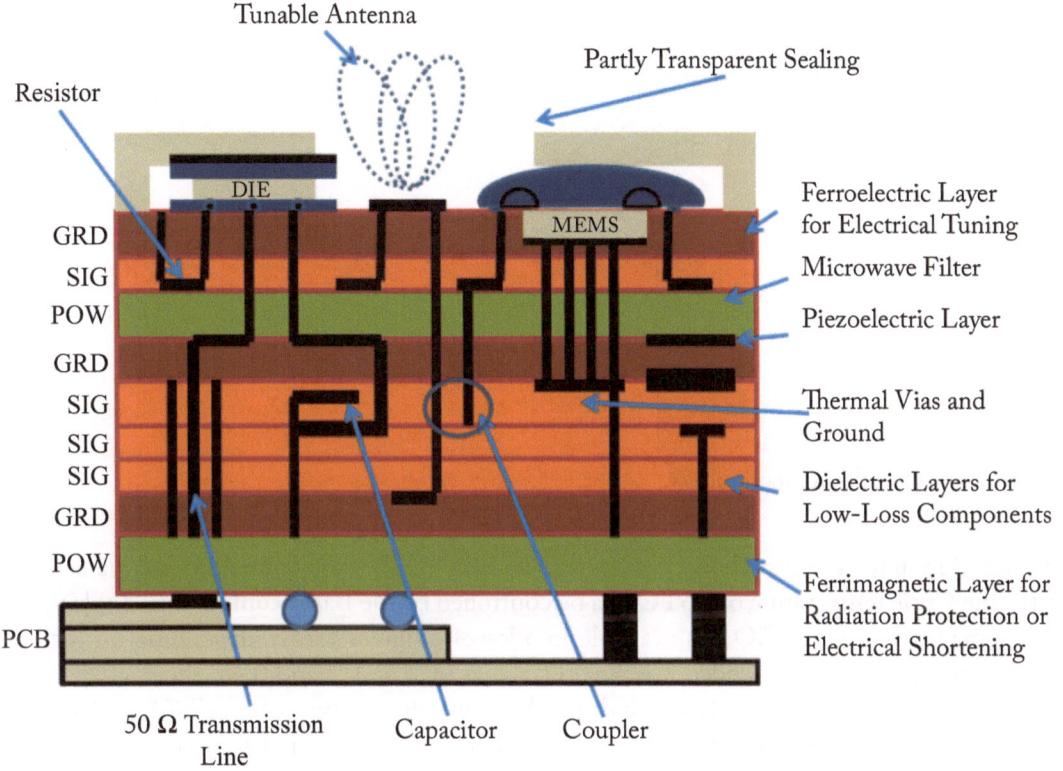

Figure 2.29: Cross-section of a typical tunable microwave circuit exploiting ferroelectric material features (LTCC modules) [40, 59, 60].

common interface must have the following [40]: (i) similar crystal symmetry, (ii) similar plane lattice parameters, (iii) similar thermal expansion coefficients, and (iv) chemical stability. In order to meet these requirements, the selection of the appropriate substrate, electrode, and dielectric materials is essential. The employment of the appropriate film growth technology and its specific parameters is also essential.

2.20   HYSTERESIS LOOP OF FERROELECTRICS

Ferroelectrics like BSTO exhibit spontaneous polarization due to a permanent dipole orientational mechanism. The dipoles are strongly coupled and form domains in a way similar to ferrites. Below the Curie temperature, they present a hysteresis loop associated with high microwave losses. Its dielectric constant can be tuned by varying the applied electric field in both its ferroelectric and paraelectric states (below and above the Curie temperature, respectively). The latter is very important since microwave losses are quite lower in the paraelectric state due to the

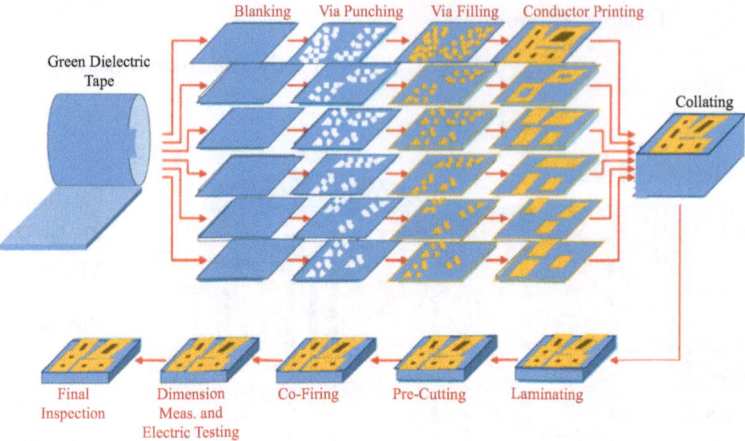

Figure 2.30: The LTCC technology.

absence of hysteresis losses. Thus, ferroelectrics like BSTO are mostly used in their paraelectric state. The Curie temperature of BSTO can be controlled by the Ba/Sr content ratio (x) [59, 61]. Thus, BSTO ($Ba_{1-x} Sr_x TiO_3$) is available in a lot of different forms, depending on the ambient temperature for its intended use. For the values of $X = 0.3$ and 0.45, the corresponding Curie temperatures are $T_C = 310$ K and 254 K, while the $0.4\,Ba_{0.55}\,Sr_{0.45}\,TiO_3 - 0.6\,MgO$ has $T_C = 209$ K. The MgO additive minimizes microwave losses but also causes a reduction in permittivity [62].

2.21   THEORY OF THE FERROELECTRIC DIELECTRIC RESPONSE

According to Tagantsev et al. [39] and the references therein, the dielectric response of ferro-electrics was given by the Ginzburg–Landau theory. This is based on a series expansion of the Helmholtz free energy F with respect to vector macroscopic polarization $\bar{P}$. When the material is considered to be isotropic, in other words, when polarization is assumed to be collinear to the macroscopic electric field, the vector notation can be dropped. Approximating free energy with the first two terms of the expansion, we have:

$$F = \frac{\alpha}{2}P^2 + \frac{\beta}{4}P^4, \qquad (2.40)$$

where α and β are constants.

The derivative of free energy with respect to polarization is equal to the electric field

$$E = \frac{\partial F}{\partial P} = \alpha P + \beta P^3. \qquad (2.41)$$

The electric permittivity (ε) of the medium is equal to the derivative of polarization P with respect to the electric field $\varepsilon = \partial P / \partial E$; in other words, observing (2.41), for any value of the electric field, permittivity is defined as the slope of the polarization curve.

$$\varepsilon = \varepsilon_0 \varepsilon_r = \frac{\partial P}{\partial E} = \frac{1}{\alpha + 3\beta P^2}. \tag{2.42}$$

Let $\varepsilon(0) = 1/(\alpha \varepsilon_0)$ or $\varepsilon_0 \varepsilon(0) = 1/\alpha$, where $\varepsilon_0 = 8.854 \times 10^{-12}$ F/m is the permittivity of free space. The dielectric constant can be written as:

$$\varepsilon_r = \varepsilon(0) \cdot \frac{1}{1 + 3\beta \cdot \varepsilon_0 \cdot \varepsilon(0) \cdot P^2}. \tag{2.43}$$

From Eq. (2.43), it is obvious that $\varepsilon(0)$ represents the relative permittivity in the absence of a biasing electric field $P = P_{dc} = 0$. According to the Ginzburg–Landau theory, the coefficient α is regarded as a linear function of temperature and vanishes at the Curie–Weiss temperature T_C.

$$\alpha = \alpha_{GL} = \frac{1}{\varepsilon_0} \frac{T - T_C}{C}, \tag{2.44}$$

where C is the Curie–Weiss constant, which, for displacive ferroelectrics, has a typical value of $C \approx 10^5$ K. This implies that the dielectric constant has high values even beyond the Curie–Weiss temperature T_C. According to [39], at $T = T_C + 200$ K Eq. (2.43) gives

$$\varepsilon(0) = \frac{1}{\varepsilon_0 \alpha} = \frac{C}{T - T_C}. \tag{2.45}$$

Therefore, $\varepsilon(0) \approx 500$.

Equation (2.44) is considered to hold for $|T - T_C|/T_C \ll 1$. However, for displacive ferroelectrics, it has been found that (2.44) is valid for temperatures up to the melting point of the material with good accuracy. For incipient ferroelectrics like $SrTiO_3$ and $KTaO_3$ (those with T_C around zero K having Debye temperature $\Theta \approx 400$ K), Eq. (2.44) has been found to apply with reasonable accuracy from 50–80 K up to the melting point of the material.

Note that the terms "displacive ferroelectric" and "quantum paraelectric" are used alternatively to the term "incipient ferroelectric." Both of them refer to ferroelectrites whose Curie–Weiss temperature is around $T_C \to 0$ K. Indicative examples include $SrTiO_3$ and $KTaO_3$. For lower temperatures, far below the Debye temperature $T \ll \Theta$, Eq. (2.44) ceases to apply and the quantum statistics of the lattice vibration must be taken into account. Among the several models proposed [39], those of Vendik et al. [63], and Barrett [64] have reasonable accuracy. The Vendik model reads as follows:

$$\alpha = \alpha_V = \frac{T_V}{\varepsilon_0 C} \left\{ \sqrt{\frac{1}{16} + \left(\frac{T}{T_V}\right)^2} - \frac{T_C}{T_V} \right\}, \tag{2.46}$$

where T_V is a parameter describing the slowing down of the temperature dependence of α. Likewise, Barrett's theory yields:

$$\alpha = \alpha_B = \frac{T_B}{\varepsilon_0 C} \left\{ \coth \left(\frac{T_B}{T} \right) - \frac{T_C}{T_B} \right\}, \tag{2.47}$$

where the T_B parameter accounts for the same effect as T_V.

2.22   FERROELECTRIC TUNABILITY

Since the variation in the dielectric constant from the applied DC biasing electric field (E_{dc}) is the basic point of interest in ferroelectrics, a quantity for evaluating different materials is required. This quantity is termed "tunability" (n) or "relative tunability" (n_r), e.g., [39]:

$$n = \frac{\varepsilon(0)}{\varepsilon_r (E_{dc})} \quad \text{or} \quad n_r = \frac{\varepsilon(0) - \varepsilon_r (E_{dc})}{\varepsilon(0)} = \frac{n-1}{n}. \tag{2.48}$$

Using in Eq. (2.43) as $\varepsilon_r = \varepsilon_r(E_{dc})$, tunability can be written as follows:

$$n = 1 + 3\beta \cdot \varepsilon_0 \cdot \varepsilon(0) \cdot P_{dc}^2. \tag{2.49}$$

In the case of weak nonlinearity $n_r \ll 1$, it is:

$$E_{dc} = \alpha P_{dc} \left(1 + \beta P_{dc}^2 \right) \approx \alpha P_{dc}. \tag{2.50}$$

In turn, tunability is approximated as follows [39]:

$$n \approx 1 + 3\beta \cdot [\varepsilon_0 \cdot \varepsilon(0)]^3 \cdot E_{dc}^2. \tag{2.51}$$

Expression (2.51) shows that for low values of E_{dc}, relative tunability is a very fast cubic function of $\varepsilon(0)$, namely, $n_r \propto \varepsilon(0)^3$, but for increasing E_{dc}, this dependence slows down. At the upper limit of E_{dc} we have:

$$\varepsilon_r(0) = \varepsilon(0) \approx 1/ \left\{ 3\varepsilon_0 \cdot \sqrt[3]{\beta E_{dc}^2} \right\} \tag{2.52}$$

and tunabilty becomes a linear function of E_{dc}, namely, $n \propto \varepsilon(0)$.

However, it is always possible to trade tunability against loss reduction and vice versa [51]. For BSTO ($Ba_X Sr_{1-X} TiO_3$) solid compositions, this compromise can be reached by properly selecting the barium percentage. Thus, a figure of merit for the evaluation of different materials should account for both tunability and loss tangent ($\tan \delta$). A quantity called "commutation quality factor K" is defined for this purpose [40]:

$$K = \frac{(n-1)^2}{n \cdot \tan \delta (E_{dc\,min}) \cdot \tan \delta (E_{dc\,max})}. \tag{2.53a}$$

Alternatively [58],

$$K = \frac{Tunability(\%)}{\tan \delta(\%)} = \frac{n(\%)}{\tan \delta(\%)} = \frac{\varepsilon_{r\,max} - \varepsilon_{r\,min}}{\varepsilon_{r\,max}} \cdot \frac{\tan \delta_{max}}{\tan \delta_{max} - \tan \delta_{min}}. \tag{2.53b}$$

2.23   FERROELECTRIC MICROWAVE LOSSES

Usually, incipient (e.g., $SrTiO_3$ or $BaTiO_3$) or conventional ferroelectrics of the displacive type (e.g., BSTO) in their paraelectric phase (above the curie temperature) are used in microwave tunable applications. Microwave losses in bulk thick or thin films are studied in the work of Guverich and Tagantsev [65], Tagantsev et al. [39], and Vendik et al. [63]. In the fundamental loss mechanism, due to interaction of the AC field with the phonons of ferroelectrics in bulk form, the term "intrinsic losses" was introduced [39, 65]. These losses are due to the absorption of energy quantum, ($hf = \hbar\omega$), when collisions of microwave photons with crystal lattice phonons occur. However, when ferroelectric films are employed, an additional loss mechanism should be included owing to the coupling of the microwave field with defects. This is referred in [39, 65] as "extrinsic losses." A detailed in-depth analysis of the loss mechanism is given in the review paper of Tagantsev et al. [39].

2.23.1   INTRINSIC LOSSES

Intrinsic losses come from the fundamental loss mechanism established for crystalline materials with a well-defined phonon spectrum. According to this theory, which is based on a quantum mechanics approach, the fundamental loss is mainly due to the absorption of quantum energy ($E_q = hf = \hbar\omega$) in collisions of microwave photons with thermal phonons, which have much higher energy. The great difference in energy causes difficulties in the satisfaction of conservation laws, and this complicated situation is described in the following three different absorption mechanisms [39]: (i) three-quantum, (ii) four-quantum, and (iii) quasi-Debye mechanism. In the three- and four-quantum mechanisms, the absorption of a field quantum ($\hbar\omega$) involves two and three phonons, respectively. Vendik et al. [63] refer to mechanisms (i) and (ii) as a "multiphonon scattering of the soft ferroelectric mode." Unlike in the quasi-Debye mechanism [63], they deal with the "transformation of microwave oscillations to acoustic ones due to scattering by regions with residual ferroelectric polarization." The total loss tangent $\tan\delta$ is the summation of $\tan\delta$ resulting from each mechanism [39, 49].

According to Tagantsev et al. [39], for tunable materials of the soft ferroelectric mode, both three- and four-quantum mechanisms yield a microwave dielectric loss:

$$\tan\delta_{ph} = \frac{\varepsilon''_{ph}}{\varepsilon} \propto \omega T^2 \varepsilon^{3/2}. \tag{2.54}$$

Equation (2.54) is valid under the following conditions:

(a) In the three-quantum mechanism, the frequency should be lower than that of the damping of the phonons ($\omega \leq r$); in other words, the frequencies of interphonon collisions must be lower than the phonon frequency in the crystal. For incipient ferroelectrics, such as $SrTiO_3$ or $KTaO_3$, the limit $\omega \leq r$ means that $f \leq 100$ GHz [39].

(b) In the four-quantum mechanism, the temperature should be high enough so that $K_B T \geq \hbar\Omega_{TO}$, which is the most common practical case. The symbol K_B is Boltzman's constant, and Ω_{TO} is the circular frequency of the lower transverse optical mode (optical phonons).

Even though the three- and four-quantum mechanisms give the same dependence of $\tan\delta$ on frequency (notice the linearity in terms of ω), temperature, and the dielectric constant, the three-quantum mechanism is dominant and yields an order of magnitude of higher losses, as can be seen in Figure 2.31.

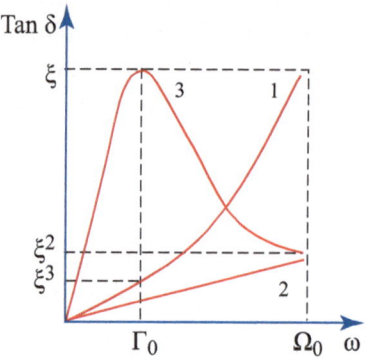

Figure 2.31: Contribution of the different mechanisms to the microwave losses of incipient ferro-electric: 1. Three-quantum, 2. Four-quantum, and 3. Quasi-Debye mechanisms. ($\xi = \Gamma_0/\Omega_0$).

It is important to note that, besides the fundamental losses, the nonlinear interaction between the soft ferroelectric mode and the thermal oscillations of the crystal lattice (multiphonon interaction) is also responsible for the ferroelectric phase transition, which reveals the high dielectric constant of the material [63].

The quasi-Debye mechanism contributes to the losses in non-centro-symmetric crystals [39]. In this case, the oscillations of the AC field modulate phonon frequencies, including a deviation from their equilibrium distribution. A relaxation of the phonon distribution, similar to the relaxation of dipole distribution, (original Debye theory), gives rise to a corresponding loss. Recall that ferroelectrics are usually used in microwave applications in their paraelectric phase, in which they exhibit a highly centro-symmetric crystalline structure. However, this is exactly true only in the absence of a DC biasing electric field, which is able to break the central symmetry and, in turn, activate the quasi-Debye loss mechanism, (DC field-induced quasi-Debye mechanism). When activated, this mechanism may represent the dominant contribution to losses for relatively low frequencies, (lower than 100 GHz) and reads as follows [39]:

$$\tan\delta_{QD}(E_{dc}) = A \cdot \omega \cdot I(E_{dc}) \cdot n_r, \qquad (2.55)$$

where n_r is the relative tunability given in Eq. (2.48).

The function $I(E_{dc}) \to 1$ for $n_r \ll 1$ and the parameter A fits the expression to experimental data (see [39] and references therein) as follows:

KTaO^3 at $T = 50$ K $\leftrightarrow A = 23 \times 10^{-3}$ CHz^{-1}, SrTiO$_3$ at $T = 80$ K $\leftrightarrow A = 17 \times 10^{-3}$ CHz^{-1}, and Ba$_{0.6}$Sr$_{0.4}$T$_i$O$_3$ at $T = 300$ K $: A = (0.7 - 1.4) \times 10^{-3}$ CHz^{-1} (calculated). The relative contribution of the Debye mechanism to the total losses is also given in Figure 2.31.

2.23.2   EXTRINSIC LOSSES

Extrinsic losses are additional losses related to defects of the material, which, due to various discontinuities (e.g., multiple layers), may be columnar or spherical inclusions. For microwave tunable applications, intrinsic and extrinsic losses are comparable. In the absence of a DC biasing field, the extrinsic contribution is dominant, whereas when bias is applied, the intrinsic loss becomes dominant [39]. The multiple layer defects are of primary importance in ferroelectric film applications. Tagantsev et al. [39] refer again to three different loss mechanisms owing to: (i) charged defects, (ii) the universal relaxation law, and (iii) the quasi-Debye contribution included by random field defects.

The research groups of Tagantsev et al. [39] and Vendik et al. [49] both seem to agree that the most important extrinsic losses in thin films are those of charged defects, so we will only focus on these.

2.23.3   LOSSES DUE TO CHARGED DEFECTS

Microwave electric field oscillations cause a motion of charged defects, which results in an acoustic wave. Hence, a part of microwave energy is transformed into acoustic wave energy, giving rise to a corresponding loss [39, 49, 63]. This type of loss plays an essential role in ferroelectric thin film technology primarily due to the Schottky defects occurring at the ferroelectric/electrode interface [39]. It is also important to note that this type of loss is proportional to the dielectric constant and, consequently, inversely proportional to the applied DC biasing electric field (assuming perovskite ferroelectrics, e.g., STO). According to Vendik et al. [49], the charged defect loss vs. frequency exhibits a wide plateau in the range of 3–30 GHz. It is interesting to observe the comparison of losses in a single crystal and a thin film sample of an STO presented in Figure 2.32 (both for zero bias), where one can verify the above statements.

2.23.4   LOSSES OF LOCAL POLAR REGIONS

Typical centro-symmetric ferroelectrics like STO exhibit local polar regions, induced by various defects and structural imperfections (though assumed to be in their paraelectric phase) [39]. For these, a quasi-Debye mechanism is expected, which causes losses as described above. This is also strongly dependent on the dielectric constant.

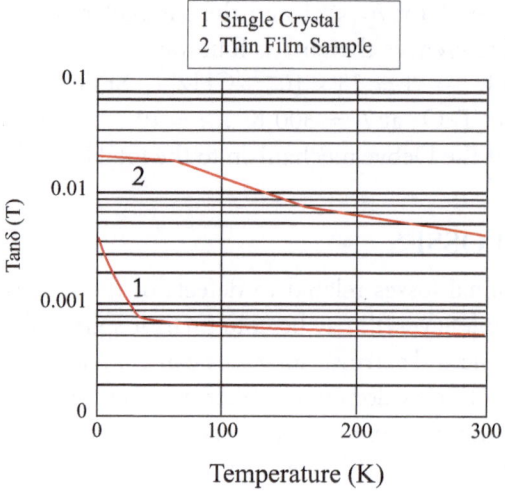

Figure 2.32: Loss tangent vs. temperature for a single crystal and a thin film sample for zero bias ($E_{dc} = 0$) at a frequency of 10 GHz.

2.24 REFERENCES

[1] WUN, Worldwide Universities Network, Advanced Materials Group, *Advances that will Revolutionize the Material World*, http://www.wun.ac.uk 15

[2] B. Norey, Twin film barium strontium titanate (BST) for a new class of tunable RF components, *Microwave Journal*, p. 210, May 2004. 15, 16

[3] I. C. Hunter and J. D. Rhodes, Electronically tunable microwave bandpass filters, *IEEE Transactions on Microwave Theory and Techniques*, vol. MTT-30, no. 9, pp. 1355–1360, September 1982. DOI: 10.1109/tmtt.1982.1131260. 15

[4] Technology Report, Microwave materials move technology forward, *High Frequency Electronics*, p. 38, May 2003. http://www.highfrequencyelectronics.com 16

[5] G. P. Rodrigue, A generation of microwave ferrite devices, *IEEE Proc.*, vol. 76, issue 2, pp. 121–137, February 1988. DOI: 10.1109/5.4389. 17, 25, 26

[6] J. L. Snoek, Magnetic and electrical properties of the binary system $MOFe_2O_3$, *Physica*, vol. 3, pp. 463–483, Elsevier, June 1936. DOI: 10.1016/s0031-8914(36)80011-1. 17

[7] C. L. Hogan, The ferromagnetic faraday effect at microwave frequencies and its applications, *Bell Systems Technical Journal*, vol. 31, pp. 1–31, January 1952. DOI: 10.1002/j.1538-7305.1952.tb01374.x. 17

[8] J. C. Sethares, Scanning the issue, guest editor of *IEEE Proc.*, special issue on *Microwave Magnetics*, vol. 76, pp. 99–102, February 1988. 17, 18

[9] D. M. Bolle and S. H. Talisa, Fundamental considerations in millimeter and near-millimeter component design employing magnetoplasmons, *IEEE Transactions on Microwave Theory and Techniques*, vol. MTT-29, no. 9, pp. 916–922, September 1981. DOI: 10.1109/tmtt.1981.1130474. 17

[10] P. DeGasperis, R. Marcelli and G. Miccoli, Magnetostatic soliton propagation at microwave frequency in magnetic garnet films, *Physical Review Letters*, vol. 59, no. 4, pp. 481–484, July 27, 1987. DOI: 10.1103/physrevlett.59.481. 17

[11] H. L. Glass, Ferrite films for microwave and millimeter wave devices, *IEEE Proc.*, vol. 76, pp. 151–158, February 1988. DOI: 10.1109/5.4391. 18, 25, 30, 31, 32, 34, 45

[12] A. J. Baden-Fuller, *Ferrites at Microwave Frequencies*, Peter Peregrinus, London, UK, 1987. DOI: 10.1049/pbew023e. xix, 18, 19, 23, 25, 26, 27, 28, 30, 33, 35, 36, 37, 38, 39, 40, 42, 45, 46

[13] R. A. Waldron, *Ferrites: An Introduction for Microwave Engineers*, D. Van Nostrand, NY, 1961. 19, 22, 23, 24, 25, 27, 35

[14] D. M. Pozar, *Microwave Engineering*, 4th ed., Wiley, NY, 2016. 19, 33, 36, 37, 38, 40, 42, 45

[15] S. Ramo, J. R. Whinnery, and T. Van Duzer, *Fields and Waves in Communication Electronics*, Wiley 1st ed. 1965, 3rd ed., NY, 1994. 22, 23

[16] A. G. Fox, S. E. Miller, and M. T. Weiss, Behavior and applications of ferrites in the microwave region, *Bell System Technical Journal*, pp. 5–101, January 1955. DOI: 10.1002/j.1538-7305.1955.tb03763.x. 24

[17] Trans-Tech, *Microwave Magnetic and Dielectric Materials*, Alpha Industries. http://www.alphaind.com/ 26

[18] Irene Hsia, Microstrip circuits and antennas in non reciprocal superstrate-substrate structure, Ph.D. Dissertation, Dept. of Electrical Engineering, UCLA, 1991. 28, 38

[19] J. D. Adam, E. Davis, G. F. Dionne, E. F. Schloemann, and S. N. Stitzer, Ferrite devices and materials, *IEEE Transactions on Microwave Theory and Techniques*, vol. MTT-50, no. 3, pp. 721–737, March 2002. DOI: 10.1109/22.989957. 28, 31, 32

[20] F. Bertaut and R. Pauthenet, Crystalline structure and magnetic properties of ferrites having the general formula $5\,Fe_2\,O_3 \cdot 3\,M_2\,O_3$, *Proc. Institute of Electric Engineers*, vol. B104, pp. 261–264, 1957. DOI: 10.1049/pi-b-1.1957.0043. 28, 30

[21] W. S. Ishak, Magnetostatic wave technology: A review, *IEEE Proc.*, vol. 76, pp. 171–174, February 1988. DOI: 10.1109/5.4393. 30, 31

[22] J. D. Adam, Analog signal processing with microwave magnetics, *IEEE Proc.*, vol. 76, pp. 159–170, February 1988. DOI: 10.1109/5.4392. 31, 33

[23] F. J. Cadieu, High coercive field and large remanent moment magnetic films with special anisotropies, *Applied Physics*, vol. 61, pp. 4105–4110, 1987. DOI: 10.1063/1.338544. 33

[24] R. A. Overfelt, C. D. Anderson, and W. L. Flanagan, Plasma sprayed $Fe_{76} Nd_{16} B_8$ permanent magnets, *Applied Physics Letters*, vol. 49, pp. 1799–1801, 1986. DOI: 10.1063/1.97195. 33

[25] B. Lax and K. J. Button, *Microwave Ferrites and Ferrimagnetics*, McGraw-Hill, NY, 1962. DOI: 10.1063/1.3051073.

[26] P. J. B. Clarricoats, *Microwave Ferrites*, Chapman & Hall, MA, 1961.

[27] R. F. Soohoo, *Microwave Magnetics*, Harper and Row, NY, 1985.

[28] D. Polder, On theory of ferromagnetic resonance, *Philosophical Magazine*, vol. 40, pp. 99–115, January 1949. DOI: 10.1080/14786444908561215. 33

[29] D. Polder, On the phenomenology of ferromagnetic resonance, *Physical Review*, vol. 73, pp. 1120–1121, May 1948. DOI: 10.1103/physrev.73.1120.3. 33

[30] L. Landau and E. Lifshitz, On the theory of dispersion of magnetic permeability in ferromagnetic bodies, *Physik Z. Sowjetunion*, vol. 8, pp. 153–169, 1935. DOI: 10.1016/b978-0-08-036364-6.50008-9. 34

[31] F. R. Morgenthaler, An overview of electromagnetic and spin angular momentum mechanical waves in ferrite media, *IEEE Proc.*, vol. 76, pp. 138–150, February 1988. DOI: 10.1109/5.4390. 35

[32] A. J. Baden Fuller, *Microwaves*, Pergamon Press, UK, 1969. 37

[33] D. M. Pozar, Radiation scattering characteristics of microstrip antennas on normally biased ferrite substrate, *IEEE Transactions on Antennas and Propagation*, vol. AP-40, pp. 1085–1092, September 1992. DOI: 10.1109/8.166534. 38

[34] I. Y. Hsia and N. G. Alexopoulos, Radiation characteristics of hertzian dipole antennas in a nonreciprocal superstrate-substrate structure, *IEEE Transactions on Antennas and Propagation*, vol. AP-40, pp. 782–790, July 1992. DOI: 10.1109/8.155743. 38

[35] G. Tyras, The permeability matrix for a ferrite medium magnetized at an arbitrary direction and its eigenvalues, *IRE Transactions Microwave Theory and Techniques*, vol. MTT-7, pp. 176–177, January 1959. DOI: 10.1109/tmtt.1959.1124645. 38, 39

[36] Y. Rahmat-Samii, Useful coordinate transformations for antenna applications, *IEEE Transactions on Antennas and Propagation*, vol. AP-27, pp. 571–574, July 1979. DOI: 10.1109/tap.1979.1142138. 38

[37] Hung-Yu Yang, A note on the mode characteristics of a ferrite slab, *IEEE Transactions on Microwave Theory and Techniques*, vol. MTT-43, pp. 235–238, January 1995. DOI: 10.1109/22.362980. 42, 44

[38] Trans-Tech application notes 655, 656, *Dielectric Properties of Ferromagnetic Materials*, Trans-Tech, Alpha Industries. http://www.alphaind.com/ xiii, 42, 44, 45

[39] A. K. Tagantsev, V. O. Sherman, K. F. Astafiev, J. Venkatesh, and N. Setter, Ferroelectric materials for microwave tunable applications, *Journal of Electroceramics*, vol. 11, pp. 5–66, Kluwer Academic Publishers, November 2003. DOI: 10.1023/b:jecr.0000015661.81386.e6. 46, 60, 61, 62, 63, 64, 65

[40] K. Khamchane, Growth and characterization of non-linear ferroelectric heterostructures, Ph.D. Thesis, Chalmers University of Technology, Göteborg, 2005. xiii, xiv, 46, 48, 49, 51, 56, 58, 59, 62

[41] Jung-Hyuk Koh, Processing and properties of ferroelectrics $Ag(Ta, Nd)O_3$ thin films, Ph.D. Thesis, Royal Institute of Technology, Stockholm, 2002. 46, 47, 48

[42] J. Valasek, Piezo-electric and allied phenomena in rochelle salt, *Physical Review*, XVII(4), pp. 475–481, 1920. DOI: 10.1103/physrev.17.475. xiv, 49, 54

[43] A. J. Baden Fuller, *Ferrites at Microwave Frequencies*, Peter Peregrinus, Exeter, UK, 1987. DOI: 10.1049/pbew023e. 49

[44] V. M. Goldschmidt, *Ber. Dtsch. Chem. Ges.*, vol. 60, p. 1270, 1927. xiv, 51, 54, 58

[45] K. Paulson, W. Breckon, and M. Pidcock, Electrode modelling in electrical impedance tomography, *SIAM Journal of Applied Mathematics*, vol. 52, no. 4, pp. 1012–1022, 1992. DOI: 10.1137/0152059.

[46] B. Noren, Thin film bariu strontium titanate (BST) for a new class of tunable RF components, *Microwave Journal*, vol. 47, no. 5, pp. 210–220, May 2004. xiv, 53, 54

[47] M. E. Lines and A. M. Glass, *Principles and Applications of Ferroelectrics and Related Materials*, Clarendon Press, Oxford, 1977. DOI: 10.1093/acprof:oso/9780198507789.001.0001.

[48] A. S. Bhalla, R. Guo, and R. Roy, The perovskite structure—a review of its role in ceramic science and technology, *Materials Research Innovations*, vol. 4, pp. 3–26, Springer Verlag, 2000. DOI: 10.1007/s100190000062. 51, 54, 55

[49] O. G. Vendik, E. K. Hollman, A. B. Kozyrev, and A. M Prudan, Ferroelectric tuning of planar and bulk microwave devices, *Superconductivity Journal*, vol. 12, no. 2, pp. 325–338, 1999. DOI: 10.1023/A:1007797131173. xiv, 55, 56, 57, 63, 65

[50] C. Ang, A. S. Bhalla, and L. E. Cross, Dielectric behaviour of paraelectric $KTaO_3$, $CaTiO_3$, and $(Ln_{1/2} Na_{1/2}) TiO_3$ under a DC electric field, *Physical Review B*, vol. 64, 184104, 2001. DOI: 10.1103/physrevb.64.184104.

[51] S. Gevorgian, O. Tageman, and A. Derneryd, Electronically scanning beam-formers based on ferroelectric technology, *Frequenz*, vol. 59, pp. 40–48, 2005. 55, 56, 62

[52] K. Uchino, K. Hosi, and S. Nomura, Ferroelectric materials and applications, *Proc. 1st Meeting in Japan*, p. 102, 1977. xiii, xiv, 51, 56, 57

[53] Aml Industry and UCLA, *Piezoelectric Materials, Background–Definition*, Active materials Lab. http://aml.seas.ucla.edu/ xiii, 51

[54] J. H. Glenn, *Perovskite Superlattices as Tunable Microwave Devices*, http://www.nasatech.com/briefs/may03/LEW16938.html 56

[55] H. M. Christen and K. S. Harshavardhan, *Perovskite Superlattices as Tunable Microwave Devices*, Glenn Research Center, NASA. http://www.nasatech.com/tsp 51, 55, 56

[56] G. A. Smolenskii Ed., *Ferroelectrics and Related Materials*, Gordon & Breach, Netherlands, 1984. 55

[57] K. Bethe, Uber das mikrowellenverhalten nichlineare dielectrica, *Philips Research Reports Supplements*, no. 2, 1970. 56

[58] O. Auciello, S. Saha, D. Y. Kaufman, S. K. Streiffer, W. Fan, B. Kabius, J. Im, and P. Baumann, Science and technology of high dielectric constant thin films and materials integration for application to high frequency devices, *Electroceramics Journal*, vol. 12, pp. 119–131, Kluwer, 2004. DOI: 10.1023/b:jecr.0000034006.59246.5e. 62

[59] Tao Hu, BST-based low temperature co-fired (LTCC) modules for microwave tunable components, Ph.D. Thesis, University of Oulu, Japan, 2004. xiv, 59, 60

[60] H. Jantunen, T. Kangasvieri, J. Vahakangas, and S. Leppavuorri, Design aspects of MW components with LTCC technique, *Journal of European Ceramic Society*, vol. 23, pp. 2541–2548, 2003. DOI: 10.1016/s0955-2219(03)00155-9. xiv, 59

[61] U. Syamaprasad, R. K. Galgali, and B. C. Mohanty, Dielectric properties of the $Ba_{1-X} Sr_X TiO_3$, *Materials Letters*, vol. 7, no. 5–6, pp. 197–200, 1988. DOI: 10.1016/0167-577x(88)90009-2. 60

[62] L. Sengupta, E. Ngo, S. Stowell, M. Oday, and R. Lancto, *Ceramic Ferroelectric Composite Material BST-MgO*, U.S. Patent 5, 427, 988, 1995. 60

[63] O. G. Vendik, L. Ter-Martirosyan, and S. P. Zubko, Microwave losses in incipient ferroelectrics as functions of the temperature and the basing field, *Journal of Applied Physics*, vol. 84, pp. 993–998, 1998. DOI: 10.1063/1.368166. 61, 63, 64, 65

[64] J. H. Barrett, Dielectric constant in perovskite type crystals, *Physical Review*, vol. 7, p. 7403, 1952. DOI: 10.1103/physrev.86.118. 61

[65] V. Gurevich and A. Tagantsev, Intrinsic dielectric loss in crystals, *Advances in Physics*, vol. 40, no. 6, pp. 719–767, 1991. DOI: 10.1080/00018739100101552. 55, 63

CHAPTER 3

Finite Ferrite Samples

3.1 DEMAGNITIZATION FACTORS AND FERRITE SAMPLES

In the analysis in Chapter 2, an infinitely extending ferrite material, biased by a DC magnetic field, was considered. In turn, the permeability tensor expressions either with or without losses are only valid for the internal fields within the ferrite and only when the alternating fields are uniform throughout all space, e.g., Hogan [1]. However, in practice, a finite ferrite specimen is used. The specimen can be in the form of small spheres, rods or posts, thin disks or plates as well as in the form of thin or thick films. Films are of particular interest in the quest for miniaturization and the integration process. Moreover, the ferrite permeability tensor is expressed in terms of the DC biasing field $\bar{H}_0 = \bar{H}_i$ internal to the ferrite. The question in the finite sample case is how this internal field $\bar{H}_i$ is related to the corresponding externally applied one $\bar{H}_a$ (which is more readily measured), that is, the DC magnetic field in the air surrounding the ferrite sample. The answer in principle is straightforward. Fields $\bar{H}_a$ and $\bar{H}_i$ are related through the boundary conditions at the ferrite-air interface. However, this is not so simple since the ferrite permeability involved depends on $\bar{H}_i$. A formulation accounting for this field should be established. A classical approach given by Kittel [2] is the introduction of demagnetization factors into the magnetic susceptibility or permeability tensor. The demagnetization factors can be defined by applying the boundary conditions at the ferrite-air interface. Let us first examine two special cases of the DC biasing field orientation with respect to the ferrite sample surface. First, we assume a planar sample (Figure 3.1a) transversely magnetized at saturation ($\bar{M} = \bar{M}_s$) as shown in Figure 3.1b. The continuity of the magnetic flux density normal to the planar sample component (B_n) yields:

$$B_n = \mu_0 H_a = \mu_0 \left(H_i + M_s \right). \tag{3.1}$$

In addition, the internal field becomes

$$H_i = H_a - M_s. \tag{3.2}$$

Equation (3.2) shows that M_s reduces the internal field with respect to its value in the air region. In contrast, when a longitudinal saturation magnetization is applied, the continuity of the tangential magnetic field $\bar{H}_t$ retains an internal magnetic field identical to the externally applied one $\bar{H}_a$, as shown in Figure 3.1c. In other words,

$$\bar{H}_t = \bar{H}_a = \bar{H}_i. \tag{3.3}$$

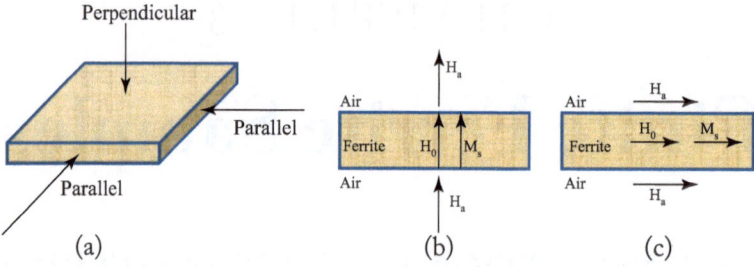

Figure 3.1: (a) Directions of internal $\bar{H}_i = \bar{H}_0$ and external $\bar{H}_a$ magnetic fields for a thin planar ferrite sample magnetized at saturation. (b) transversely—normal and (c) longitudinally—tangential.

In general, the internal DC magnetic field normal to the sample surface is reduced by the projection of the saturation magnetization along the direction of each component. For such an arbitrary bias orientation, the two continuity conditions (3.2) and (3.3) can be rewritten as follows:

$$\text{Normal components:} \quad \hat{n} \cdot \bar{H}_i = \hat{n} \cdot \left(\bar{H}_a - \bar{M}_s \right) \tag{3.4}$$

$$\text{Tangential components:} \quad \hat{n} \times \bar{H}_i = \hat{n} \times \bar{H}_a, \tag{3.5}$$

where $\hat{n}$ is the outward unit vector normal to the sample surface of the ferrite.

When an AC external magnetic field is also applied, this gives rise to an AC magnetization $\bar{M}$. This magnetization causes a reduction in the corresponding internal AC field component normal to the sample surface. By symbolizing the total (including DC) internal and external magnetic fields with $\bar{H}_i$, $\bar{H}_e$, respectively, and the total magnetization with $\bar{M} = \bar{M}_{ac} + \bar{M}_s$, the internal magnetic field can be written as [3, 4],

$$\bar{H}_i = \bar{H}_e - [N]\bar{M}, \tag{3.6}$$

where $[N] = [N_x, N_y, N_z]$ is the demagnetization factor. According to the above analysis, the elements N_x, N_y, and N_z are obviously dependent on the sample's shape and are independent of the DC bias direction (this only defines the direction or components of $\bar{M}_s$). By definition, we have,

$$N_x + N_y + N_z = 1. \tag{3.7}$$

The demagnetization factors are accurately calculated only for ferrite samples of an ellipsoidal shape and are given in [4]. Note that for all other (non-ellipsoidal) shapes, magnetization is non-uniform, and hence the macroscopic field varies from point to point within the sample. Therefore, there is no macroscopic field that can be defined for the sample as a whole [1]. Suitable ellipsoidal shapes approximate practically useful samples such as rods and disks. Their demagnetization factors can be defined by providing that the sample dimensions are small compared to the wavelength. These are (see Figure 3.2):

$$\text{Rod along the } z\text{-axis:} \quad N_x = N_y = 1/2, \ N_z = 0 \tag{3.8a}$$

$$\text{Sphere:} \quad N_x = N_y = N_z = 1/3 \tag{3.8b}$$

Thin disk or plate with

$$\text{a perpendicular } z\text{-axis:} \quad N_x = N_y = 0, \ N_z = 1. \tag{3.8c}$$

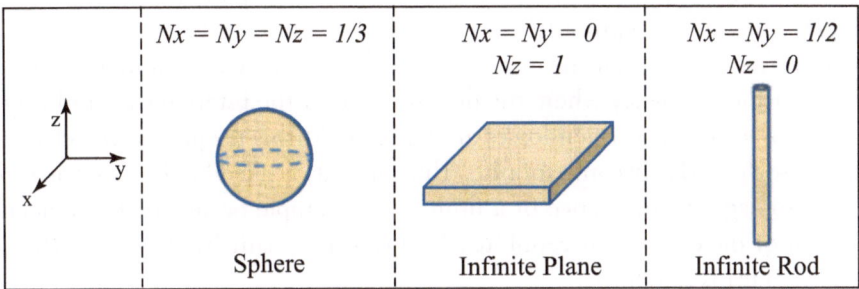

Figure 3.2: Elements N_x, N_y, and N_z of three different shapes of ferrite samples.

Returning to Eq. (3.6), magnetization $\bar{M}$ is related to the internal magnetic field $\bar{H}_i$ through the magnetic susceptibility $\mathcal{X}$ previously given from Eqs. (2.15), (2.18), and (2.19) as:

$$\bar{M} = [\mathcal{X}]\bar{H}_i. \tag{3.9}$$

By substituting $\bar{H}_i$ from Eq. (3.6) into Eq. (3.9), a new relation is obtained between $\bar{M}$ and the magnetic field outside the sample at an infinitesimal distance from its surface. This relation is

$$\bar{M} = [\mathcal{X}]\bar{H}_e - [\mathcal{X}][N]\bar{M}$$

or

$$\{[I] + [\mathcal{X}][N]\} \ \bar{M} = [\mathcal{X}]\bar{H}_e. \tag{3.10}$$

Equation (3.10) can be written in the form $\bar{M} = [\mathcal{X}_e]\bar{H}_e$, where $[\mathcal{X}_e]$ is called "the external susceptibility tensor." That is because $\mathcal{X}_e$ expresses magnetization in terms of the field just outside the sample (for details, see [3] or [4]). To examine the effects of the demagnetization phenomenon on the gyromagnetic resonance, let us repeat the external susceptibility expression for the simple case in which the DC bias is applied along the z-axis [3]:

$$M_x = \mathcal{X}^e_{xx} H_{xe} + \mathcal{X}^e_{xy} H_{ye} = \frac{\mathcal{X}_{xx}\left(1 + \mathcal{X}_{yy} N_y\right) - \mathcal{X}_{xy} \cdot \mathcal{X}_{yx} N_y}{D} H_{xe} + \frac{\mathcal{X}_{xy}}{D} H_{ye} \tag{3.11a}$$

$$M_y = \mathcal{X}^e_{yx} H_{xe} + \mathcal{X}^e_{yy} H_{ye} = \frac{\mathcal{X}_{yx}}{D} H_{xe} + \frac{\mathcal{X}_{yy}\left(1 + \mathcal{X}_{xx} N_x\right) - \mathcal{X}_{yx} \cdot \mathcal{X}_{xy} N_x}{D} H_{ye} \tag{3.11b}$$

$$M_z = 0 \tag{3.11c}$$

$$D = \left(1 + \mathcal{X}_{xx} N_x\right) \cdot \left(1 + \mathcal{X}_{yy} N_y\right) - \mathcal{X}_{yx} \cdot \mathcal{X}_{xy} \cdot N_x \cdot N_y. \tag{3.11d}$$

Note that expressions (3.11a), (3.11b), (3.11c), and (3.11d) relate the external AC magnetic field (outside the sample) to the corresponding AC components of the magnetization. Futhermore, the internal susceptibility expression $[\mathcal{X}]$ is already given in Eq. (2.17). The important point to stress here is that the gyromagnetic resonance of an infinitely extending ferrite medium occurs at the frequency where the denominator of the internal susceptibility $[\mathcal{X}]$ vanishes, that is, at $\omega_r = \omega = \omega_0$, when losses are ignored. At this frequency, the elements of the permeability tensor μ and k become infinite. Thus, it is also called "Polder's resonance" [5]. In contrast, the gyromagnetic resonance of a finite ferrite sample occurs at the frequency where the denominator of the external susceptibility $[\mathcal{X}_e]$ vanishes, namely, at $D = 0$. Thus, ignoring losses, Eq. (3.11d) reads as follows:

$$\left(1 + \frac{\omega_0 \omega_m N_x}{\omega_0^2 - \omega^2}\right)\left(1 + \frac{\omega_0 \omega_m N_y}{\omega_0^2 - \omega^2}\right) - \frac{\omega^2 \omega_m^2 \cdot N_x N_y}{\left(\omega_0^2 - \omega^2\right)^2} = 0. \tag{3.12}$$

Solving Eq. (3.12) for the gyromagnetic resonance frequency of a z-biased sample, we have:

$$\omega_r = \omega = \sqrt{\left(\omega_0 + \omega_m N_x\right)\left(\omega_0 + \omega_m N_y\right)}. \tag{3.13}$$

In (3.13), we substitute frequencies $\omega_m = \mu_0 \gamma M_s$ and $\omega_0 = \mu_0 \gamma H_0$, where $H_0 = H_{iz} = H_a - N_z M_s$ is the internal DC biasing field in the z-direction. An expression, known as "Kittel's equation," [1, 2, 5, 6] results for the resonance frequency called "Kittel's resonance":

$$\omega_r = \mu_0 \gamma \sqrt{\left[H_a + (N_x - N_z) M_s\right] \cdot \left[H_a + (N_y - N_z) M_s\right]} = \mu_0 \gamma H_{\mathit{eff}}. \tag{3.14}$$

It is important to realize from Eq. (3.14) that the gyromagnetic resonance of a finite sample depends on its demagnetization factors (N_x, N_y, and N_z) and, in turn, on its shape. The appearance of H_{eff} and, consequently, the dependence of Kittel's resonance on saturation magnetization (M_s) are also shown in Eq. (3.14). Recall that at the gyromagnetic resonance of a lossless ferrite, magnetic susceptibility becomes infinite, driving magnetization to infinity. However, when losses are taken into account, the above quantities are finite but take their maximum value. As stated by Hogan [1], a serious confusion may result when considering what actually happens in a finite ferrite sample at the two resonance frequencies $\omega_r = \omega_0 = \mu_0 \gamma H_0$ and

$\omega_r = \mu_0 \gamma H_{eff}$. A physical reasoning may prove very helpful in understanding these phenomena. When the circular frequency (ω) of the applied AC field is far below ω_0, the AC magnetization components are very small and there is a very small difference between the magnitudes of internal $\bar{H}_i$ and external $\bar{H}_e$ magnetic fields. As frequency ω is increased, approaching ω_0 (that is, $\omega \approx \omega_0$), magnetization reaches infinity (maximized in the lossy case) and, following Eq. (3.6), the internal field $\bar{H}_i$ tends to zero (see Hogan [1], or Waldron [5]). This Polder's condition is rather physically paradoxical or presents a mathematical singularity. That is because the internal field $\bar{H}_i$ generates magnetization through internal magnetic susceptibility as $\bar{M} = [\mathcal{X}]\bar{H}_i$. Therefore, when $\mathcal{X}$ tends to infinity and $\bar{H}_i$ tends to zero, a singularity occurs and magnetization $\bar{M}$ is indefinable. According to Waldron [5], in the Polder's condition, the microwave energy is excluded from the ferrite specimen and magnetization also remains small. From a different point of view, this situation can be understood through the boundary conditions as follows. The magnetic field source outside the specimen acts on it, producing a very high permeability at $\omega = \omega_0$, which, in turn, reflects this AC field or does not allow any AC energy penetration. When losses are considered, the internal field to be built is just enough to make up for the energy loss of the electron spins' precessional motion [1].

Returning to Kittel's resonance, as the frequency of the applied field is increased above $\omega_0 (\omega > \omega_0)$, the phase of magnetization ($\bar{M}$) reverses with respect to the externally applied field $\bar{H}_e$. If we also consider the minus sign in Eq. (3.6), then the magnitude of the internal field starts to increase until it reaches its maximum value at Kittel's resonance, $\omega_r = \mu_0 \gamma H_{eff}$, given in Eq. (3.14) [1]. Under this condition, the internal field ($\bar{H}_i$) becomes theoretically infinite (ignoring losses) or practically maximum (including losses), and magnetization $\bar{M}$ is maximized as well. However, under Kittel's condition, the large magnetization observed is not due to the gyromagnetic resonance, but due to the concentration of microwave energy inside the specimen [5]. In view of the above analysis, microwave losses under Polder's resonance condition are small due to the negligible energy concentration inside the sample. In contrast, at Kittel's resonance there is heavy absorption (high losses) of microwave energy caused by the large fields in the specimen [5].

3.2   SPIN WAVES AND MAGNETOSTATIC WAVES

Up to this point, we have seen cases in which a ferrite medium was placed in a uniform magnetic field strong enough to produce saturation. The exchange field forces act to align all magnetic dipoles parallel to one another, and all are lined up with the applied DC field. It was then considered that the whole ferrite medium behaves as a large magnetic dipole precessing about the DC magnetic field. From a different point of view, all the individual spins precess in phase. When an additional uniform, circularly polarized (CP), high-frequency magnetic field acts on the ferrite, the spins' precession angle is increased or decreased depending on the circular right- (RHCP) or left (LHCP)-handed polarization. Furthermore, for the above uniform field, all spins still precess in phase. If the field is non-uniform, for example if it is applied to one part of a ferrite sample, then the precession angle will be increased (assume an RHCP field) in the area where this field is applied. The internal exchange fields will tend to align their neighboring spin dipoles; in other words, they will act in such a way as to swing their neighbors into the largest precession angle [7]. The above will happen *with some delay*. In this manner, the largest precession angle disturbance will travel through the crystal lattice in the form of a wave. This is called a "spin-wave" and appears to have both phase and amplitude to change from dipole to dipole [8]. If a high-frequency field with LHCP is applied to one part of the ferrite sample, the disturbance would be a smaller precession angle traveling again through the sample as a spin-wave. The wavelength of low-order spin-wave modes can be very long and can be expected to occur in the very low microwave range, even as "magnetostatic waves." However, high-order spin waves may have very short wavelengths. In any case, it is important to know how these wavelengths are compared with the dimensions of the ferrite specimen. When the wavelength is quite larger than the sample dimensions, propagation effects may be ignored, and the corresponding waves are "magnetostatic waves." In contrast, when the wavelength is comparable or shorter than the ferrite specimen, propagation must be taken into account, and the waves are "spin waves."

To generalize the above example, a spatially non-uniform high-frequency magnetic field can potentially excite spin waves. This non-uniform magnetic field could come from the corresponding externally applied magnetic field. Two mechanisms are usually responsible for this internal non-uniformity and, as a result, for the excitation of spin waves. The first one concerns the ferrite's non-linear behavior when high-power microwave fields are applied to it [4, 5]. The second mechanism concerns different anisotropy fields often caused by geometrical irregularities [7, 8].

Starting with the first mechanism, when the microwave magnetic field exceeds a certain critical value, the magnetization in the direction of the biasing DC field (e.g., M_z for rods and disks, where z is the axis of symmetry) becomes unstable. Assume, for example, a transversely biased disk with ($\bar{H}_{DC} = H_0 \hat{z}$) for which the magnetization component $M_z = M_s + m_z$ (m_z is the microwave part) decreases locally, perhaps due to thermal vibration. This, in turn, corresponds to an increase in angle-θ between the electrons' spin axes and the $\hat{z}$-direction of the polarizing field. If M_z is stable, it will increase, by a reduction of θ, to its initial value. In contrast, if M_z is unstable, it will continue to decrease (increase of θ) until a new steady state is reached [5]. This local disturbance of spin precessional motion will affect the neighboring electron spins through exchange forces. In this manner, the disturbance will propagate in all directions within the ferrite specimen as a spin-wave. Something worth keeping in mind is that saturation magnetization (M_s) is a major part of M_z. Thus, the characteristics of spin waves are expected to depend strongly on M_s as well as on the exchange forces. In addition, this phenomenon may be accompanied by strong demagnetization.

The second mechanism occurs mostly in normally biased thin ferrimagnetic films. In this case, a uniform microwave field may excite long wavelength (i.e., magnetostatic) spin waves. Actually, the electron spins on the film's surface observe different anisotropy fields than those of the electron spins within the film [7, 8]. The application of an external microwave field excites spin-wave modes, which may only have an odd number of half wavelengths across the film's thickness. Modes with an even number of half wavelengths (even order) cannot be excited since they have no net energy interaction with the film. To understand this, recall that the even-order modes tend to present zero (or minimum) field intensity at the film's surface, where the potentially unstable electron spins lie. In contrast, the odd-order modes present maximum field intensity at the film's surface. Once again, note that these are long-wavelength spin waves.

3.3   LOW– VS. HIGH–ORDER SPIN WAVES

In a series of experimental investigations reviewed by Walker [9], the ferrite sample was placed in a metallic cavity at a point where the RF magnetic field was sufficiently heterogeneous, yielding an effective excitation of spin waves. These investigations revealed multiple resonances (spin-wave modes), in addition to the expected Kittel resonance. "Dimensional" resonances were also expected. However, for small samples with dimensions of the order of a few mils (1 mil $= 0.001'' = 0.0254$ mm $= 25.4$ μm), these resonances were avoided. The observed resonance frequencies were *substantially independent of sample size, markedly dependent upon sample shape*, and also *dependent upon saturation magnetization* [9]. Likewise, when a ferrite sample of finite size is immersed in a heterogeneous microwave field, standing waves will be set up in the same modes. These phenomena could be explained with the aid of a model consisting of an array of dipoles, which directly interact with a uniform DC magnetic field and mutually interact through exchange of electromagnetic forces. Exchange interactions are of a short range. When a substantial change in the direction of transverse magnetization occurs, the exchange magnetic field at a distance X (cm) is of the order of $(10^{-8}/X^2)$ oersted. Furthermore, *a force is important only when its range is comparable with the wavelength*. In order to be comparable with the commonly applied external magnetic field, (a few kilogauss), the exchange field must have the mode wavelengths of about $\lambda_g \approx 10^{-5}$ cm. Such wavelengths give resonances of some hundreds of THz for plane or TEM waves where $\lambda_g = \lambda_0/\sqrt{\varepsilon_r \mu_r}$ [10]. Likewise, for resonances in the GHz range, a field of a few gauss may be obtained. Thus, for low-order spin-wave modes (magnetostatic modes), the exchange field is negligible. As the mode order is increased, its wavelength is decreased, and the exchange field becomes increasingly important. Up to the very high-order spin waves, the exchange field prevails, and since demagnetizing fields become negligible, these waves (e.g., their resonances) do not depend on the specimen's shape [5]. In fact, their wavelength is so short that spin waves may be analyzed as plane waves even within a small ferrite sample [4].

To conclude this comparison, we can say that spin waves are dominated by exchange fields and, in contrast, are negligible for magnetostatic waves. However, both waves are strongly affected by the external magnetic field. Thus, exchange fields must be included in the wave equation so that we can obtain the characteristic equation of spin-wave modes. Therefore, in their analysis, the size as well as the shape of the specimen can be ignored. Consequently, demagnetization factors will be also ignored.

3.4   MAGNETOSTATIC MODES

As described above, there are low-order spin wave modes, which present long wavelengths. Due to the long wavelength, the "exchange forces" are negligible and each spin performs a precessional motion. This motion stems from two kinds of fields. One is the external magnetic field (both DC biasing and microwave field), modified by the demagnetization factors. The other is the dipolar field produced by the other spins; in other words, each spin generates a field, which acts on the precessional motion of all other spins and vice versa. This is usually modeled as a dipole-dipole interaction of one spin with all the others. In turn, the produced modes depend on the shape of the specimen since the shape defines the demagnetization factors (N). It is important to realize that this dipole-dipole interaction is the mechanism that creates the energy that acts to demagnetize the sample in order to avoid free poles at the surface [10–12]. For this reason, this energy is often obtained from the solution of a magnetostatic boundary value problem. In other words, this is the same mechanism that creates the "demagnetization factors N." Moreover, propagation is important when the size of the ferrimagnetic specimen is comparable with the wavelength. The most commonly used specimens with ellipsoidal shape have dimensions of a few mils and are very small compared to the wavelengths of the modes. Therefore, for such cases, the mode spectrum may be considered independent of specimen size and propagation may be ignored in their actual magnetostatic analysis. However, there are a lot of practical applications whose dimensions are not small enough to neglect propagation. Propagation yields dependence from the sample size.

To sum up, the magnetostatic mode characteristics can be obtained from the solution of a magnetostatic wave equation $\nabla \times \bar{H} \approx 0$ by taking the demagnetization factors (N) into account. Therefore, these modes strongly depend on the shape of the sample, which, in turn, defines the demagnetization factors. Also, recall that exchange fields will be ignored.

Due to the dependence of magnetostatic modes on sample shape, a lot of researchers contributed to this area. Walker [10] was the first who studied the magnetostatic modes for spheres, and these are often called "Walker modes." The most important modern applications of magnetostatic modes involve ferrite thin films or slabs. Damon and Eshbach [13] first proposed the corresponding theory.

3.5   SPIN–WAVE SPECTRUM MANIFOLD

The spin-wave spectrum is composed of high-order modes excited whenever a heterogeneous internal magnetic field occurs in a ferrimagnetic sample. Their wavelength is primarily determined by the exchange field and, secondarily, by the external applied magnetic field while it is independent of demagnetizing fields. Due to the latter, spin waves are shape independent, while due to their short wavelength, they are also independent of sample size. In fact, their wavelength is so small compared to practical specimen dimensions that spin waves can be analyzed as plane waves even within a small ferrite sample [4]. Spin waves as well as magnetostatic waves absorb energy from the applied magnetic field and heat up the ferrite.

Our primary task in this section is the study of the "spin-wave dispersion equation," which will serve as a tool for the determination of their spectrum, also known as "spin-wave manifold." The latter will also serve as a reference point for the determination of the magnetostatic wave spectrum. As explained above, the key to spin-wave analysis is the consideration of exchange fields. This is not a trivial task. On the contrary, an accurate analysis requires a quantum-mechanical approach. Therefore, the analysis below focuses on an intuitive physical understanding rather than a mathematically rigorous one, which can be found in the cited references [12] and [14–38].

3.6 EXCHANGE–FIELD INTERACTION

Exchange energy has an electrostatic origin, which stems from electron wavefunctions (space and spin together) interacting toward the minimization of Coulomb energy [6, 12, 14, 15]. Depending upon the material, exchange interaction is due to the electronic orbital overlap of neighboring atoms, either direct or mediated by conduction electrons (see [14] and references therein). The latter is called "RKKY (Ruderman–Kittel–Kasuya–Yosida) interaction" or "indirect exchange."

There is another type of interaction called "super exchange," where the electronic orbital overlap is mediated by intervening non-magnetic ions. From a quantum-mechanics point of view, exchange coupling is a consequence of the Pauli exclusion principle, which states that "two electrons cannot occupy the same quantum state at the same place and time." Here the "spin up" and "spin down" quantum states are considered. The consequence of this principle is that two parallel spins cannot overlap. Overlapping can occur only when two spins are antiparallel. However, in this manner, the Pauli exclusion principle keeps parallel spins apart, which, in turn, lowers their electrostatic energy. Their exchange energy is equal to the amount by which the Coulomb energy is reduced.

In general, the exchange energy density W_{ex}, can be written in a form also known as "Heisenberg Hamiltonian" (e.g., [12, 14]):

$$W_{ex} = -\frac{1}{V} \sum_{<m,n>} J_{mn} \bar{S}_m \cdot \bar{S}_n, \tag{3.15}$$

where V is the volume of the material, $\bar{S}_m$ and $\bar{S}_n$ are the atomic spins of the mth and nth atoms, and J_{mn} is a quantum-mechanical coefficient known as "the exchange internal." Equation (3.15) has a discrete form and, in general, assumes a summation of the exchange energy involved in the interaction of all spin pairs within the sample volume. It is helpful to give the following form for the exchange energy of any (m, n) pair [15]:

$$W_{mn} = -J_{mn} \bar{S}_m \cdot \bar{S}_n = -J_{mn} \left(\frac{\bar{\mu}_m}{g\mu_B} \right) \cdot \left(\frac{\bar{\mu}_n}{g\mu_B} \right) = -\frac{J_{mn}}{g^2 \mu_B^2} \bar{\mu}_m \cdot \bar{\mu}_n. \tag{3.16}$$

In the above equation, the atomic spins $\bar{S}_m$ and $\bar{S}_n$ are actually expressed in terms of the atomic magnetic moments $\bar{\mu}_m$ and $\bar{\mu}_n$ through the well-known expression

$$\bar{\mu}_i = g\mu_B \bar{S}_i, \tag{3.17}$$

where g is the Lande factor and μ_B is the Bohr magneton.

Micromagnetic calculations in the discrete form of expression (3.15) are too complicated for real samples since only calculations up to a few atoms are feasible. For this purpose, only macro-magnetic (macroscopic) theories are capable of handling the problem. Toward this direction, the discrete exchange is approximated by a continuous function (often accounting for the six nearest neighboring spins) and exchange energy density is expressed in terms of the local magnetization components (M_x, M_y, and M_z) [12–25] as:

$$W_{ex} = A \cdot \left\{ (\nabla m_x)^2 + (\nabla m_y)^2 + (\nabla m_z)^2 \right\} = A \cdot \frac{|\nabla \bar{M}|^2}{|M|^2}, \tag{3.18a}$$

where $\hat{m} = \bar{M}/|\bar{M}|$ is the unit vector along the local magnetization direction, and A denotes the exchange or stiffness constant given in general by:

$$A = \frac{J \cdot S^2}{\alpha} K_{neighbor}. \tag{3.18b}$$

For a cubic crystal and six neighboring spins, it is:

$$A = \frac{J}{2\mu_B^2 g^2}. \tag{3.18c}$$

The expressions assume that only the nearest $K_{neighbor}$ atoms interact, and this number depends on the crystalline structure of the material ($K_{neighbor} = 1$ for cubic, 2 for body centered cubic (BCC), 4 for face centered cubic (FCC), and $2\sqrt{2}$ for hexagonal close-packed (HCP)). Moreover, α is the lattice constant (mean distance between atoms). The neighboring spins are assumed to be aligned in parallel, $\bar{S}_m = \bar{S}_n = \bar{S} = \bar{\mu}/g\mu_B$ having the same exchange integral $J_{mn} = J$. Volume V in (3.15) is approximated as $V \approx \alpha^3$, while the spin summation yields an α^2 in the numerator. Likewise, the spin direction in (3.18a) is assumed to be identical to that of local magnetization, i.e., $\bar{S} = \hat{m}|\bar{S}|$.

The exchange interaction can be represented by an effective magnetic field ($\bar{H}_{eff}^{ex}$). Since magnetic energy can be written as an integral of $(1/2\bar{M} \cdot \bar{H})$, the effective field can be defined by the total derivative of energy variation [24]:

$$\bar{H}_{ex}^e = -\frac{\delta W_{ex}}{\delta M} = \frac{2A}{|M|} \nabla^2 \hat{m}(\bar{r}) = -H_e \alpha^2 \frac{\nabla^2 \bar{M}(\bar{r},t)}{|\bar{M}|}, \tag{3.19a}$$

where

$$H_e = 2\frac{JS^2}{|M|\alpha^3} = 2\frac{A}{|M|\alpha^2}.$$

(3.19b)

The other energy term, which significantly affects spin waves, is anisotropy energy, since this is a source producing internal field in homogeneity and turns spin waves.

3.7   ANISOTROPY ENERGY

The term "magnetic anisotropy" refers to a direction of "easy" or preferential magnetization. When spins are aligned along this direction, magnetic energy becomes minimum (local minimum). In general, the term "anisotropy" refers to phenomena of different energy density for magnetization in different directions. Anisotropy can be an inherent magneto-crystalline or artificially induced property. The spin-orbit coupling in the relativistic interaction may produce magneto-crystalline anisotropy. This anisotropy is greatly affected by crystal lattice symmetry. Any change in its parameters, including those due to local temperature, will modify it. Artificially induced anisotropy may be due to the strain during growth or directional growth. In some sandwiched thin layers, strain-based anisotropy can create an out-of-plane anisotropy directly competing with shape anisotropy [16].

Besides its bulk component, magneto-crystalline anisotropy also has a surface component called "surface anisotropy," which is important in thin films. On surfaces and interfaces where lattice symmetry is broken, surface anisotropy can be enhanced to an order of magnitude even greater than that of bulk crystalline anisotropy [16].

It must be noted at this point that "demagnetization energy" is usually categorized as a form of shape anisotropy, which has already been considered in previous sections through the demagnetization factors [N]. This is the result of long-range magnetic dipole interaction, and it is strongly affected by sample boundaries. This definition may cause serious confusion since shape anisotropy may be attributed to the dipole-dipole interaction of one spin with all the others. However, magneto-crystalline anisotropy could also be studied as a dipole-dipole interaction. For this reason, many researchers studying spin waves refer to dipole-dipole interaction as the second-most important (after exchange energy) source of spin-wave excitation. It seems that the key to distinguishing between mechanisms exciting spin waves and those exciting magnetostatic waves is the range of interaction. Therefore, short-range interactions generate short wavelength spin modes, while long-range interactions generate long-wavelength magnetostatic modes. Following this reasoning, along with the explanation given in previous sections, shape anisotropy or the corresponding demagnetization field does not affect spin waves. However, the analysis of magnetostatic modes through a solution of a magnetostatic boundary value problem is of primary importance.

Returning, we will see the energy ($W_{anis,u}$) for a uniaxial anisotropy, that is, for a crystal with only one energetically favorable axis. Consider, for example, a hexagonal crystal where magnetization is oriented at an angle θ with respect to the c-axis of symmetry (or $\hat{z}$-axis), as shown in Figure 3.3.

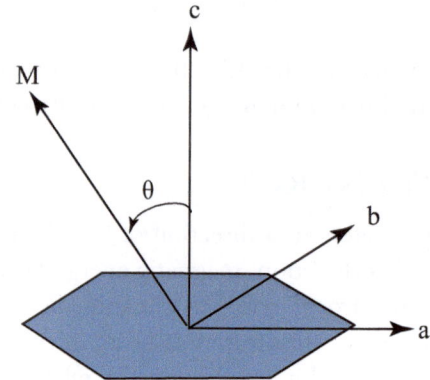

Figure 3.3: Hexagonal crystal where magnetization is oriented at an angle θ with respect to the axis of symmetry.

The uniaxial anisotropic energy can be written as follows [24]:

$$W_{anis,u} \approx -k_1 \cos^2 \theta + k_2 \cos^4 \theta = -k_1 m_z^2 + k_2 m_z^4. \tag{3.20a}$$

In most hexagonal crystals, the c-axis is the axis of easy magnetization, ($|k_2| \ll |k_1|$). Thus [24]:

$$W_{anis,u} = K_u \sin^2 \theta, \tag{3.20b}$$

where K_u is the uniaxial anisotropy constant and θ is the angle of the observation point with respect to the anisotropy axis.

According to the classical approach given by Wolf [18], anisotropy energy depends on the magnitudes and directions of the individual sub-lattice normalized magnetic moments ($\hat{m}_i$). It usually takes the form of a sum over sub-lattices:

$$W_{anis} = \sum_{ij} F_i \left(\hat{m}_i, \hat{m}_j \right). \tag{3.21a}$$

Equation (3.21a) is approximated by a power series, which is convergent, provided that local anisotropy does not exceed or is not comparable with exchange energy. For a cubic symmetry (usually encountered in ferrites), Eq. (3.21a) reads as follows [12, 18, 24]:

$$W_{anis} = const + K_1 \left(m_x^2 m_y^2 + m_x^2 m_z^2 + m_y^2 m_z^2 \right) + K_2 m_x^2 m_y^2 m_z^2, \tag{3.21b}$$

where K_1 and K_2 are the anisotropy constants ($k_1 > 0$ for Fe and $K_1 < 0$ for Ni). Likewise, for a hexagonal symmetry, occurring for example in hexaferrites, the power series expansion of

(3.21a) takes the form:

$$W_{anis} = const + K'_1 \sin^2 \theta + K'_2 \sin^4 \theta + K'_3 \sin^6 \theta + K'_4 \sin^6 \theta \cdot \cos(6\varphi) + \dots \quad \text{(3.21c)}$$

where θ is the angle between $\bar{M}$ and the crystal axis and φ defines the direction of its projection at $\theta = \pi/2$ plane.

In turn, anisotropy effects can be included in the magnetization, or LLG (Landau–Lifshitz–Gilbert), equation through an equivalently applied magnetic field ($\bar{H}_{anis}$) [16]:

$$\bar{H}_{anis} = -\frac{1}{|M|} \nabla W_{anis} \left(\bar{M} \right), \quad \text{(3.22a)}$$

where $\bar{H}_{anis}$ can be approximated by [12]:

$$H_{anis} = \frac{2K}{M_s}, \quad \text{or} \quad \bar{H}_{anis} = \frac{2K}{M_S^2} M_z \hat{z}. \quad \text{(3.22b)}$$

Note that for YIG, the anisotropy constant K is negative (see [30] and references therein).

Finally, note that anisotropy energy represents a mechanism essentially due to the crystalline properties of the material. Thus, for a polycrystalline ceramic with randomly oriented crystallites, the average anisotropy energy per unit volume is zero, yielding $\bar{H}_{anis} = 0$ [17].

3.8   MAGNETIZATION EQUATION FOR SPIN WAVES

The basic tool used to characterize spin waves as well as any other type of waves in ferrites is the magnetization equation. This equation must be in a proper form to include the above described energy terms exciting spin waves. In fact, this is the LLG equation given above. It should also be pointed out that for spin waves, the exchange interaction and anisotropy energy terms will be considered while the magnetostatic or demagnetization field will be ignored. Let us rewrite the magnetization equation:

$$\frac{d\bar{M}}{dt} = -\gamma \mu_0 \bar{M} \times \left(\bar{H} + \bar{H}^m \right), \quad \text{(3.23a)}$$

where

$$\bar{H}^m = \bar{H}_{anis} + \bar{H}_{ex}. \quad \text{(3.23b)}$$

Note that $\bar{H} = \bar{H}_{DC} + \bar{H}_{AC}$ is the externally applied magnetic field, which includes both DC biasing and the microwave field components. Besides that, microwave losses are not accounted for in (3.23a) for convenience. Recall that losses can be considered by adding a term $\bar{H}_{loss}$ in (3.23b). Moreover, it is worth pointing out that the contribution of the material to the magnetic field $\bar{H}^m$ does not belong to the Maxwell equations [17]. In general, (3.23a) is non-linear since $\bar{H}^m$ depends on $\bar{M}$ and only a linearized version can be solved. This can be obtained through approximations, like those of Eqs. (3.20), which remove $\bar{H}^m$ dependence from $\bar{M}$.

Instead of solving the LLG magnetization equation, spin waves can also be obtained from the Holstein–Primakoff Hamiltonian [19]. For this purpose, spin waves will be treated as quasi-particles called "magnons" by direct analogy with phonons, which represent mechanical waves through crystal ferrites. In this case, even some similarities to photons may be identified.

3.9   SPIN WAVES AS MAGNONS

The energy exciting spin waves is directly proportional to atomic quantities and, in turn, electron spins are quantized. Based on this, "magnons" are defined as quasi-particles of flipped magnetization [12]. The quantized energy of spin waves can be written as [18]:

$$E_K = \hbar \cdot \omega_k = \frac{h}{2\pi} \cdot 2\pi f_k = h \cdot f_k, \tag{3.24}$$

where h is the Plank constant and ω_k is the spin-wave circular frequency.

Note that since the energy is quantized, spin waves exist only at some definite frequencies $\omega = \omega_k = \omega_k(k)$, depending on their wavenumber k. This will be shown to be equal to the wave's resonance frequency. It can be proven [20, 21] that the exchange energy of each spin wave ($W_{ex,k}$) varies proportionally to its wavenumber k as $W_{ex,k} = Dk^2$ or

$$W_{ex,k} = \left(\frac{2\gamma \hbar A}{M_s} \right) \cdot k^2 = D \cdot k^2. \tag{3.25}$$

The exchange constant D in CGS is $D = 2A/M_s$, while in SI, as in (3.25), is $D = 2\gamma \hbar A/M_s$. According to Wolf [18], D can be roughly estimated by $D \approx \alpha^2 K_B T_N$, where α is the mean lattice spacing, K_B is the Boltzman constant, and T_N is the *Néel temperature*. For YIG, it is $D \approx 5 \cdot 10^{-9}$ Oe $\cdot$ cm^2 [22]. The total energy of the system can be employed to find a quantized classical Heisenberg Hamiltonian [12]:

$$H = \sum_K \left(D \cdot k^2 + \gamma \hbar H_0 \right). \tag{3.26a}$$

Holstein–Primakoff also included the dipole-dipole contribution to yield a Hamiltonian of the form

$$H = \sum_K (\hbar \omega_K) \alpha_K \cdot \alpha_K^*, \tag{3.26b}$$

where α_K^* and α_K are the creation and destruction operators defined in [19] and the original references therein.

3.10 SPIN WAVES IN AN INFINITE MEDIUM

Let us consider an infinite ferrite medium biased at saturation by a DC magnetic field H_0 in the z-direction. In addition, assume that exchange interaction is the dominant mechanism exciting spin waves, that is, neglecting anisotropy. This could be a realistic approximation for some bulk ferrites, but not for thin films. In turn, the LLG equation (3.23) must be solved with $\bar{H}_{ex}$ given by (3.19). In addition, recall that *for very large sample dimensions, spin waves may be regarded as plane waves*. Thus, the magnetization vector magnitude could be considered equal to its saturation value perturbed by an infinite sum of spin waves. Therefore, the denominator of (3.19) reads $|\bar{M}| \approx M_s$. Moreover, a Fourier series expansion can be considered for $\bar{M}(\bar{r}, t)$ [16, 23]:

$$\bar{M}(\bar{r}, t) = \bar{M}_s + \sum_{K \neq 0} \bar{m}_K(\bar{r}, t) = \bar{M}_s + \sum_{K \neq 0} \bar{m}_K e^{+j\bar{k}\cdot\bar{r}} e^{j\omega t}, \qquad (3.27)$$

where $\bar{m}_K = (m_{Kx}, m_{Ky}, m_{Kz})^T$, $\bar{k} \cdot \bar{r} = k_x x + k_y y + k_z z$,

$$k_x = k \sin\theta_K \cos\varphi_K, \quad k_y = k \sin\theta_K \sin\varphi_K, \quad \text{and} \quad k_z = k \cos\theta_K.$$

At this point, it can be readily proved that the exchange energy is proportional to k^2 or $W_{ex} = Dk^2$. For this purpose, consider a single spin wave as a plane wave proportional to $\propto e^{+j\bar{k}\cdot\bar{r}} e^{-j\omega t}$, for which it can be easily proved that $\partial/\partial x = +jk_x$, $\partial/\partial y = +jk_y$, $\partial/\partial z = +jk_z$ and, in turn, $\bar{\nabla} = +j\bar{k}$ or $(\bar{\nabla})^2 = -k^2$. Using (3.18), we have W_{ex} proportional to Dk^2.

Following Herring and Kittel [25], and aiming at the resonance frequency of each individual spin wave (magnon), we may seek solutions of (3.23) in the form of

$$\bar{M} = \bar{M}_s + \bar{m}_K e^{j\bar{k}\cdot\bar{r}} e^{-j\omega_K t}. \qquad (3.28)$$

The corresponding exchange magnetic field of (3.19) becomes

$$\bar{H}_{ex} = + \left(\frac{2A}{M_s}\right)(-k^2)\frac{\bar{m}_K}{M_s} = -Dk^2\frac{\bar{m}_K}{M_s}. \qquad (3.29)$$

It is assumed that the magnetization magnitude remains almost constant and equal to its saturation value $|\bar{M}| \approx M_s$, and only its direction changes over time. Considering DC bias in the $\hat{z}$-direction, $\bar{H}_{DC} = H_0\hat{z}$ and $\bar{M}_s = M_s\hat{z}$, the LLG equation (3.23) can be used as a layout for Cartesian coordinates adopting the usual small signal approximation $|\bar{H}_{ac}| \ll |\bar{H}_{DC}|$, which yields $M_z = M_s + m_{kz} \approx M_s$ and $H_z = H_0 + H_{zac} \approx H_0$. For example, we may write:

$$\bar{m}_K = [m_{Kx}, m_{Ky}, 0] \quad \text{and} \quad \bar{M} = [m_{Kx}, m_{Ky}, M_s]. \qquad (3.30)$$

Instead of solving the LLG equation once again, it is easier to make it equivalent to the alread-solved demagnetization case. Observing (3.26) in comparison with (3.29), the exchange field may be included in the LLG equation (3.23) as an equivalent term $[N_{eq}][M]$ provided that $N_{eqz} = 0$ since (3.29) does not include $\bar{M}_s = M_s \hat{z}$ and $N_{eqx} = N_{eqy} = N_{eq} = Dk^2/M_s$. In turn, the corresponding magnetic susceptibility is given by (3.9), and the spin-wave resonance frequency is given by (3.13), which reads as follows:

$$\omega_K = \sqrt{(\omega_0 + \omega_m N_{eqx}) + (\omega_0 + \omega_m N_{eqy})}$$
$$= \left[\mu_0 \gamma H_0 + (\mu_0 \gamma \cdot M_s) \cdot Dk^2/M_s \right] = \mu_0 \gamma \left(H_0 + Dk^2 \right). \tag{3.31a}$$

We must keep in mind that (3.30) does not include anisotropy effects or the dipole-dipole interaction. Recall that from (3.22b), the anisotropy field is approximately constant in time, and for a $\hat{z}$-biased ferrite, it is also $\hat{z}$-directed, $\bar{H}_{anis} = (2k/M_s)\hat{z}$ [26]; that is, it is parallel to $\bar{H}_{DC} = H_0 \hat{z}$ and it can be directly added to it, yielding a modification in ω_0 as

$$\omega_0 \rightarrow \omega_{0\alpha} = \gamma\mu_0 \left(H_0 + H_{anis} \right) = \gamma\mu_0 \left(H_0 + 2k/M_s \right). \tag{3.31b}$$

For the above approximation to be valid, that is, to ignore dipole-dipole interaction or, equivalently, ignore the demagnetization fields, the wavelength of the spin wave λ_n must be at least a few times smaller than the sample dimensions (ℓ_S). An approximate limit [12] to ignore the boundary conditions at the sample air interface is that $|k| > k_{min}$ and $k_{min} = 2\pi/\lambda_{n\,min} \approx 10/\ell_S$. According to [12], above this limit, the pole (magnetic charges ρ_m) distribution induced by $\bar{m}_K e^{j\bar{k}\cdot\bar{r}}$ at the sample-air interface alternates signs very rapidly. In turn, the field created by these poles (dipole-dipole interaction) does not reach an appreciable distance inside the sample, (vanishing at negligible distance from the interface). It can in turn be considered a zero magnetic charge density or $\nabla \cdot \bar{M} = 0$. Requiring (3.27) to have zero divergence yields:

$$\nabla \cdot \bar{M} = j \left(k_x \hat{x} + k_y \hat{y} + k_z \hat{z} \right) \cdot \bar{m}_k = 0 \leftrightarrow k_x = k_y = 0. \tag{3.32}$$

It is obvious that, due to symmetry, it is required to have $k_x = k_y = 0$, and the gyroscopic precession cone is circular. A representative spin wave is given in Figure 3.4 [20].

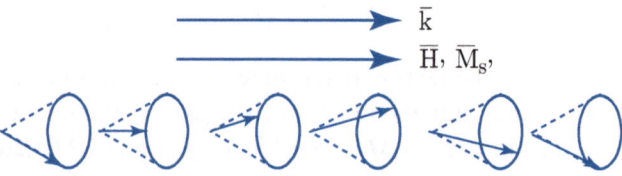

Figure 3.4: A spin wave with ignored dipolar contributions and $\bar{k}/\bar{M}_s$ [20].

3.11 SPIN WAVES INCLUDING DIPOLAR INTERACTIONS

It has been noted in the previous sections that spin waves have a short wavelength and can be treated as plane waves affected only by exchange and anisotropy fields. Moreover, they are independent of demagnetization fields. At the end of the spectrum, magnetostatic modes are independent of exchange interactions, which are short-range forces and dependent mostly on demagnetization fields. However, there is a gray area between these two states, which is defined for $0^+ < k < k_{min}$.

According to Suhl [23], spin waves with wavelengths comparable to sample dimensions are not normal modes of the system even to first order. In order to clarify this, a short overview of the basic principles of demagnetization energy will be given. Starting from the divergence of the magnetic flux density $\bar{B} = \mu_0(\bar{H} + \bar{M})$ inside the sample, an equivalent magnetic charge density, (or pole density), $\rho_m(\bar{r})$ can be defined. From

$$\nabla \cdot \bar{B}(\bar{r}) = 0 \leftrightarrow \nabla \cdot \mu_0 \left(\bar{H}(\bar{r}) + \bar{M}(\bar{r}) \right) = 0 \tag{3.33a}$$

we have

$$\nabla \cdot \bar{H}(\bar{r}) = -\nabla \cdot \bar{M}(\bar{r}) = \rho_m(\bar{r}) \quad \text{or} \quad \rho_m = -\nabla \cdot \bar{M}. \tag{3.33b}$$

Assuming a perfectly lossless ferrimagnetic material, free electric charges do not exist. Thus, the electric current density can be set equal to zero $J = 0$. Furthermore, the frequency of these spin waves may be considered low enough to ignore the displacement current ($\varepsilon \frac{\partial \bar{E}}{\partial t} = j\omega\varepsilon\bar{E}$), at least as far as the generation of the magnetic field is concerned. Therefore, following [24, 25] and [27], the magnetic field curl can be set equal to zero for the induced dipolar field ($\bar{H}_d$), which can then be expressed in terms of a static magnetic potential $\Phi_m(\bar{r})$:

$$\nabla \times \bar{H}_d = 0 \leftrightarrow \bar{H}_d(\bar{r}) = -\nabla\Phi_m(\bar{r}). \tag{3.34}$$

Substituting back (3.33) reads as follows:

$$\nabla^2\Phi_m(\bar{r}) = -\rho_m(\bar{r}). \tag{3.35}$$

Moreover, in addition to the volume magnetic charge density $\rho_m(\bar{r})$, the discontinuity of magnetization at the sample-air interface produces an *effective magnetic surface-charge density* $\sigma_m(\bar{r})$. This can be easily expressed in terms of $\bar{M}(\bar{r})$ by imposing the boundary condition, (integrating $\nabla \cdot \bar{M}$), at the interface [27],

$$\sigma_m(\bar{r}) = \hat{n}(\bar{r}) \cdot \bar{M}(\bar{r}), \tag{3.36}$$

where $\hat{n}(\bar{r})$ is the outward unit normal at any surface point $(\bar{r})$.

The resulting magnetic potential is also given in [27] as

$$\Phi_m(\bar{r}) = -\frac{1}{4\pi} \iiint\limits_{v} \frac{\nabla' \cdot \bar{M}(\bar{r}')}{|\bar{r} - \bar{r}'|} \, dv' + \frac{1}{4\pi} \oiint\limits_{s} \frac{\hat{n}(\bar{r}') \cdot \bar{M}(\bar{r}')}{|\bar{r} - \bar{r}'|} ds'. \tag{3.37}$$

It is important to note that when magnetization can be considered uniform through specimen volume v, the first term of (3.37) vanishes since $\nabla \cdot \bar{M}$ becomes zero. In contrast, the surface-charge density term always exists for finite samples; it is what we have already called "shape anisotropy," accounting for the demagnetization factors $[N]$. This is exactly the case in which uniform magnetization is considered and shape anisotropy is included in Eqs. (3.1) through (3.12). Moreover, the analysis of the previous section and hence Eq. (3.31), is valid for a uniform magnetization since it does not account for any magnetic-charge densities. Always remember that isolated magnetic charges (monopoles) cannot exist, so they always appear as pairs of their opposite.

A qualitative explanation of the physical phenomena involved in the demagnetization field may prove valuable in understanding spin and magnetostatic waves. To begin with, let us define magnetization energy [24] as:

$$W_d(\bar{r}) = -\frac{1}{2} \iiint_v \bar{M} \cdot \bar{H}_d \, dv = \frac{[\mathcal{X}_e]}{2} \iiint_v |\bar{H}_d|^2 \, dv \geq 0. \tag{3.38}$$

It is clear that W_d is always positive, but the basic principle of energy minimization requires that the demagnetizing field distributes itself in such a way as to avoid the creation of volume as well as surface magnetic charge densities. The above is known as the "*pole avoidance principle.*" Concerning volume charge, it is avoided when magnetization is uniformly ($\nabla \cdot \bar{M} = 0$) aligned in a specific direction. In contrast, the avoidance of surface poles requires that $\bar{M}$ must be parallel to the specimen surface to achieve ($\hat{n} \cdot \bar{M} = 0$). It is obvious that there is a contradiction in these two tendencies since the fulfillment of one excludes the other. Thus, demagnetization energy cannot be equal to zero. A compromise of the two requirements of minimizing W_d will be achieved, as shown, for example, in Figure 3.5 [24].

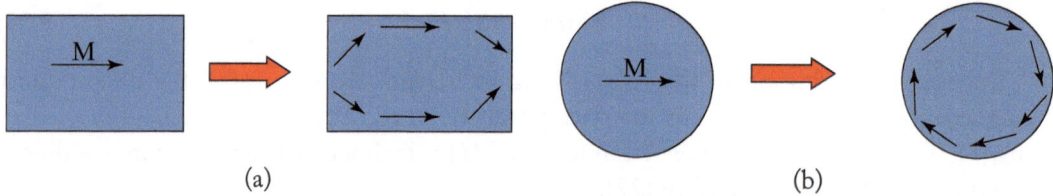

(a)   (b)

Figure 3.5: The pole avoidance principle causes demagnetization energy minimization by aligning the magnetization parallel to the surface and maintaining it as uniform as possible in the interior: (a) parallelepiped and (b) spherical specimen [24].

From the above, and particularly from Figure 3.5, it is obvious that demagnetization is a long-range mechanism since it affects practically the whole specimen volume and its surface. However, the above consideration is accurate only at magnetostatics, e.g., when frequency is low enough where $\nabla \times \bar{H} = 0$ is valid, or the specimen's dimensions are so small that this mechanism becomes strong or even dominant.

Formation of domain walls: Besides demagnetization, exchange-energy minimization strives to align all spins or keep magnetization uniform, acting in the same direction as volume-pole avoidance and contrary to surface-poles avoidance. This contradiction becomes stronger in the neighborhood of the specimen's surface. The overall energy minimization compromise yields the *division into chunks*, which are *domains with uniform magnetization* and different direction, in each domain. Thus, energy is minimized within each domain. Moreover, energy between domains is also minimized by forming flux loops, as shown in Figure 3.6 [24].

Figure 3.6: The formation of domains minimizes exchange and demagnetization energy [24].

However, the formation of domain walls causes an increase in exchange energy. Their final formation depends on whether this energy increase is lower than the corresponding reduction in demagnetization energy. Furthermore, a finite length is required for the change of the direction of magnetization between neighboring domains. For a very small specimen, the increase in energy for the formation of domain walls may be quite higher than the reduction in demagnetization energy. This results in a single domain or a vortex state as shown in Figure 3.7 [24]. For ferrimagnetic films, the primary effect of the demagnetizing field is to make the direction of magnetization point parallel to the plane of the film because of surface poles. However, demagnetization due to surface poles may be ignored in the interior of the film. Even though there are long-range effects, they do decay with distance, eventually becoming negligible [24].

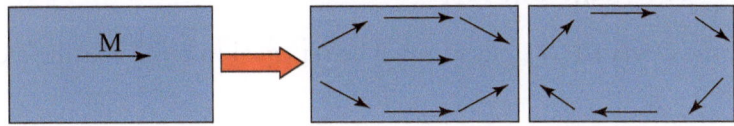

Figure 3.7: Single domain or a vortex state in very small specimens [24].

Returning to low-order spin waves, their wavelength is usually large enough to ignore surface-pole demagnetization effects, but pole volumes must be considered. This is usually termed "dipole-dipole interaction."

3.12   SPIN–WAVES ACCOUNTING FOR DIPOLE–DIPOLE INTERACTION

Considering only the volume poles $\nabla \cdot \bar{M}$ and ignoring the surface poles (to be accounted for magnetostatic modes), the dipole-dipole interaction field ($\bar{H}_d$) can be evaluated from Eq. (3.34).

Assuming a plane-wave spin-mode magnetization of the form of (3.28), the dipolar field reads [25–28]:

$$\bar{H}_{dip,k} = \bar{H}_{d0} \cdot e^{j\bar{k}\cdot\bar{r}} e^{-j\omega t}, \tag{3.39a}$$

with

$$\bar{H}_{d0} = -\frac{(\bar{k} \cdot \bar{m}_k)\,\bar{k}}{k^2}. \tag{3.39b}$$

Equations (3.39) constitute an important step in the analysis since they provide us with a field distribution all over the sample. However, they are only valid for the assumption of plane waves. Note that the difference of (3.39) from the corresponding references is due to the use herein of the SI system of units instead of the CGS. In addition, $\bar{m}_k$ is transverse to $\bar{M}_s$ since any small signal component along the DC bias direction has no effect at all. The dipolar magnetic field of (3.39) should now be included in the LLG equation (3.23) through $\bar{H}_m$ as:

$$\bar{H}_m = \bar{H}_{anis} + \bar{H}_{ex} + \bar{H}_{dip}. \tag{3.40}$$

Since $\bar{H}_{dip}$, like $\bar{H}_{ex}$, is proportional to M, it seems possible to handle the solution of the LLG equation by adding terms to some equivalent $[N]_{eq}$ factors. In such a case, the magnetic susceptibility may once again be obtained from (3.9) and the spin-wave resonance frequency from (3.11). However, there is an additional difficulty in this case. A closer observation of (3.39) reveals that $\bar{H}_{dip,k}$ can be related to $\bar{m}_k$ only through a tensor $[N]_{eq}$. As in the previous case, the ferrite will be assumed to be magnetized in the z-direction, namely:

$$\bar{H}_{DC} = H_0\hat{z}, \quad \bar{M}_s = M_s\hat{z}, \quad \bar{m}_k = \left[m_{kx}, \ m_{ky}, \ 0\right]^T \quad \text{and}$$
$$\bar{M} = \left[m_{kx}, \ m_{ky}, \ M_s\right]^T. \tag{3.41}$$

Therefore, the $\hat{z}$-related $[N]_{eq}$ factors will be set equal to zero. Starting from (3.39), this can be written as:

$$\bar{H}_d = -\hat{x}\left(\frac{k_x^2}{k^2}m_{kx} + \frac{k_xk_y}{k^2}m_{ky}\right) - \hat{y}\left(\frac{k_xk_y}{k^2}m_{kx} + \frac{k_y^2}{k^2}m_{ky}\right) \quad \text{or}$$
$$\begin{bmatrix} H_{dx} \\ H_{dy} \end{bmatrix} = -\begin{bmatrix} N_{xx}^d & N_{xy}^d \\ N_{yx}^d & N_{yy}^d \end{bmatrix}\begin{bmatrix} m_{kx} \\ m_{ky} \end{bmatrix}. \tag{3.42}$$

For convenience, let us first ignore H_{anis}, which at any time can be added to the external biasing field H_0 by a modification of ω_0 as in (3.31). Also, remember that losses can always be included by making ω_0 complex. In turn, the tensor $[N]_{eq}$ will describe the exchange $\bar{H}_e$ and the dipolar fields, to which the former contributes only its main diagonal. Rewriting (3.10) for this case, we have:

$$\bar{M} = [\mathcal{X}]\bar{H}_e - [\mathcal{X}][N]_{eq}\bar{M}, \tag{3.43}$$

where $[\mathcal{X}]$ is:

$$[\mathcal{X}] = \begin{bmatrix} \mathcal{X}_{xx} & \mathcal{X}_{xy} & 0 \\ \mathcal{X}_{yx} & \mathcal{X}_{yy} & 0 \\ 0 & 0 & 0 \end{bmatrix} \quad \text{and} \quad \begin{aligned} \mathcal{X}_{xx} &= \mathcal{X}_{yy} = \frac{\omega_0 \omega_m}{\omega_0^2 - \omega^2} \\ \mathcal{X}_{xy} &= -\mathcal{X}_{yx} = \frac{j\omega\omega_m}{\omega_0^2 - \omega^2}. \end{aligned} \tag{3.44}$$

Likewise, the equivalent N-factors read as follows:

$$[N]_{eq} = \begin{bmatrix} N_{xx} & N_{xy} & 0 \\ N_{yx} & N_{yy} & 0 \\ 0 & 0 & 0 \end{bmatrix}, \quad \begin{aligned} N_{xx} &= \left(Dk^2/M_s\right) + \left(k_x^2/k^2\right) \\ N_{yy} &= \left(Dk^2/M_s\right) + \left(k_y^2/k^2\right) \\ N_{xy} &= N_{yx} = k_x k_y/k^2. \end{aligned} \tag{3.45}$$

It is worth solving (3.43) for the common small signal approximation in its more general form. Then, it can be simplified by substituting (3.44) and (3.45) since this does not involve any extreme complexity. For this purpose, we lay (3.43) as a linear system for $\bar{M}$-components as unknowns to be solved in terms of $\bar{H}_e$-components. After some algebra, we end up with:

$$M_x = \frac{1}{\Delta} \left\{ \left(\mathcal{X}_{xx} + N_{yy} \cdot \mathcal{X}_\Delta\right) H_{xe} + \left(\mathcal{X}_{xy} - N_{xy} \cdot \mathcal{X}_\Delta\right) H_{ye} \right\} \tag{3.46a}$$

$$M_y = \frac{1}{\Delta} \left\{ \left(\mathcal{X}_{yx} - N_{yx} \cdot \mathcal{X}_\Delta\right) H_{xe} + \left(\mathcal{X}_{yy} + N_{xx} \cdot \mathcal{X}_\Delta\right) H_{ye} \right\} \tag{3.46b}$$

$$\Delta = 1 + N_{xx}\mathcal{X}_{xx} + N_{yy}\mathcal{X}_{yy} + N_{xy}\mathcal{X}_{yx} + N_{yx}\mathcal{X}_{xy} + N_\Delta \mathcal{X}_\Delta, \quad \text{where} \tag{3.46c}$$

$$\mathcal{X}_\Delta = \mathcal{X}_{xx}\mathcal{X}_{yy} - \mathcal{X}_{xy}\mathcal{X}_{yx} \quad \text{and} \quad N_\Delta = N_{xx}N_{yy} - N_{xy}N_{yx}. \tag{3.46d}$$

Note that expressions (3.46) reduce exactly to that of (3.11) when $N_{xy} = N_{yx} = 0$. Moreover, (3.46) can be considered for the definition of a more general susceptibility tensor. The most interesting part of (3.46c) is the spin-wave resonant frequency, which is given by setting $\Delta = 0$. Exploiting (3.44), this reads:

$$\Delta = 1 + \mathcal{X}_{xx}\left(N_{xx} + N_{yy}\right) + \mathcal{X}_{xy}\left(N_{yx} - N_{xy}\right) + N_\Delta \mathcal{X}_\Delta, \tag{3.47}$$

where $\mathcal{X}_\Delta$ is simplified to:

$$\mathcal{X}_\Delta = \omega_m^2/\left(\omega_0^2 - \omega^2\right). \tag{3.48}$$

Substituting to (3.45) we have:

$$N_{xx} + N_{yy} = 2\frac{Dk^2}{M_s} + \sin^2\theta_k \tag{3.49a}$$

$$N_\Delta = \frac{Dk^2}{M_s}\left(\frac{Dk^2}{M_s} + \sin^2\theta_k\right), \tag{3.49b}$$

where θ_k is the angle of wavenumber $\bar{k}$ with respect to the z-axis, along which $\bar{M}_s$ is oriented.

The spin-wave characteristic equation now reads as follows:

$$\Delta = 1 + \frac{Dk^2}{M_s} \mathcal{X}_{xx} + \left(\frac{Dk^2}{M_s} + \sin^2 \theta_k \right) \cdot \left(\mathcal{X}_{xx} + \frac{Dk^2}{M_s} \mathcal{X}_\Delta \right). \tag{3.50}$$

Substituting the $\mathcal{X}$-components from (3.44) and $\mathcal{X}_\Delta$ from (3.48) while letting $\Delta = 0$, we have:

$$\Delta = 0 \leftrightarrow 0 = \left(\omega_0^2 - \omega^2 \right) + \frac{Dk^2}{M_s} \omega_0 \omega_m + \left(\frac{Dk^2}{M_s} + \sin^2 \theta_k \right) \cdot \left(\omega_0 \omega_m + \frac{Dk^2}{M_s} \omega_m^2 \right). \tag{3.51}$$

It is then relatively easy to solve for the spin-wave resonance frequency:

$$\omega = \omega_k = \left[\left(\omega_0 + \frac{Dk^2}{M_s} \omega_m \right) \left(\omega_0 + \frac{Dk^2}{M_s} \omega_m + \omega_m \sin^2 \theta_k \right) \right]^{1/2}. \tag{3.52a}$$

In order to get the familiar expression appearing in most of the relevant references, let us substitute for $\omega_m = \mu_0 \gamma M_s$ to yield:

$$\omega_k = \left[\left(\omega_0 + \mu_0 \gamma Dk^2 \right) \left(\omega_0 + \mu_0 \gamma Dk^2 + \mu_0 \gamma M_s \sin^2 \theta_k \right) \right]^{1/2}. \tag{3.52b}$$

Expressions (3.52) are identical to the original ones given by Herring and Kittel in [25], even though they used the CGS system of units and some different symbols. It must be noted that the authors in [25] have directly substituted the exchange and dipolar fields in the LLG equation. The resulting system of equations was directly solved for ω. Herein, for clarification reasons and to be compatible with the previous sections, we thought it would be preferable to solve it for the susceptibility tensor and obtain the spin-wave frequency from the resonance condition $\Delta = 0$. To account for the ferrite anisotropy field $H_{anis} = 2k/M_s$, one may substitute ω_0 in (3.52b) as $\omega_0 \rightarrow \omega_{0a}$ according to (3.31), and the spin-wave resonant frequency will read as follows:

$$\omega_k = \mu_0 \gamma \left[\left(H_0 + H_{anis} + Dk^2 \right) \cdot \left(H_0 + H_{anis} + Dk^2 + M_s \cdot \sin^2 \theta_k \right) \right]^{1/2}. \tag{3.52c}$$

3.13   SPIN–WAVE MANIFOLD

The spin-wave manifold, or its dispersion diagram $(\omega$–$k)$, can be obtained as a graphic representation of (3.52), as shown in Figure 3.8.

Figure 3.8 and Eq. (3.52c) show that the possible resonance frequencies for each specific wavenumber lie between a minimum $\omega_k (\theta_k = 0°)$ and a maximum $\omega_k (\theta_k = 90°)$ value (branches).

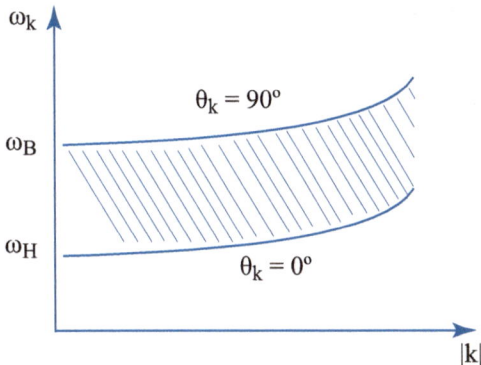

Figure 3.8: Spin-wave manifold [12, 18, 20]. The shaded area denotes the magnetostatic wave spectrum.

For the lower branch, $\theta_k = 0°$, Eq. (3.52) reduces to (3.31) and, thus, corresponds to the case in which only exchange interaction is significant, while the volume demagnetization magnetic charges (poles) vanish ($\nabla \cdot \bar{M} = 0$). Equivalently, magnetization is uniform throughout the sample. This also yields a wavenumber and a wave propagation parallel to the saturation magnetization ($\hat{z}$-axis), as already explained in Figure 3.3 and Eqs. (3.20). That happens because $\bar{k} = k_z\hat{z}$, $k_x = k_y = 0$, and Eq. (3.45) gives $N_{xx} = N_{yy} = D \cdot k^2/M_s$ and $N_{xy} = N_{yx} = 0$ (zero dipolar contribution). Inserting these into (3.46) yields $M_x = M_y$, which verifies that the gyroscopic precession cone for $\theta_k = 0°$ is circular.

In contrast, for the upper branch, $\theta_k = 90°$, the dipolar effects are maximized, and the wavenumber is perpendicular to the saturation magnetization or the DC biasing field. That means that $\bar{k} = k_x\hat{x} + k_y\hat{y}$, $k_z = 0$. A visual presentation of spin waves for $\theta_k = 90°$ is given in Figure 3.9.

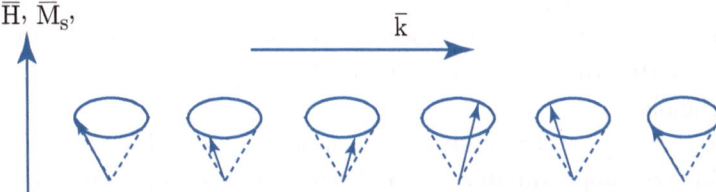

Figure 3.9: A visual presentation of spin waves for maximum dipolar interaction, where $\theta_k = 90°$ [20].

Note that in this case the non-zero equivalent N-factors make $M_x \neq M_y$ in (3.46) and, in turn, the precession cone becomes elliptical. A more detailed picture of the spin-wave manifold for YIG is shown in Figure 3.10. This has been adapted from Buffler [28], and appeared also in [18].

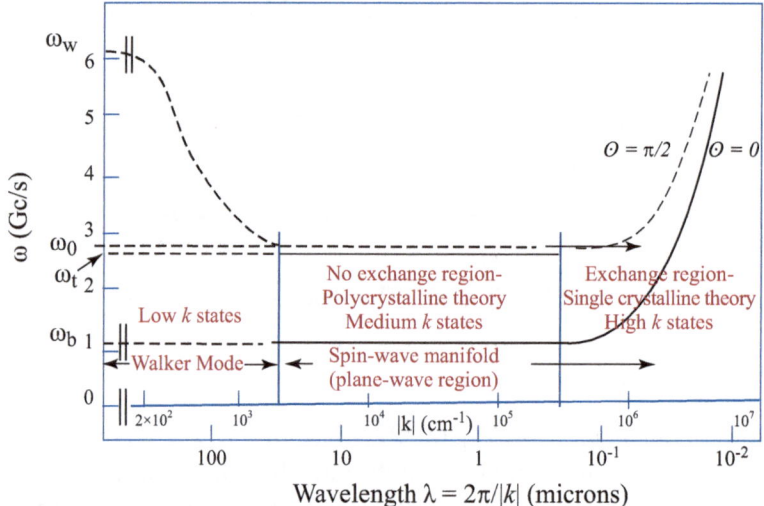

Figure 3.10: Spin-wave manifold for YIG including polycrystalline theories without exchange interaction and magnetostatic waves [28].

As previously mentioned, the spin-wave manifold does not extend to very low wavenumbers, where the wavelength becomes comparable with the sample-size dimensions. In this case, the surface magnetic charge becomes dominant and must be accounted for through the demagnetization factors, while the wavelength is so large that exchange interaction may be ignored. Here, we will give certain values of the lower limit k_{min}. Suhl [23] mentioned an approximate figure of $k_{min} = 2\pi/\lambda_{max} \approx 10/(\text{sample dimensions})$ or $\lambda_{max} \approx 0.6/(\text{sample dimensions})$. According to Patton [20], it is $k_{min} \approx 10^2$ to 10^5 (cm^{-1}), depending on the dimensions of the sample. For a typical YIG sample, dimensions of the order of $\sim 400\mu$m $= 0.4$ mm [28] designate the region 2×10^3 cm$^{-1} < k < 4 \times 10^5$ cm^{-1} as a spin-wave region for which exchange interaction is negligible.

Besides the lower spin wavenumber limit, an upper limit also exists. According to Patton [20], the spin-wave approximation is valid when the maximum wavenumber is $k_{max} = 2\pi/\lambda_{min} << \pi/\mathbf{a}$. Quantity $\mathbf{a}$ is the crystal lattice constant and k_{max} is smaller than the Brillouin zone edge. This could also be similar to the Nyquist sampling rate since $\mathbf{a} >> \lambda_{min}/2$; in other words, spin waves are set up by a number of successive atomic spins, so there should be an adequate number of atomic spins within half a wavelength.

3.14   PRELIMINARIES TO SPIN–WAVE EXCITATION

Before we go into the analysis of magnetostatic waves [30], it is worth discussing the excitation of spin waves (a subject to be revisited later) since it has been a point of disagreement which is of primary interest. In the early days of ferrimagnetism, researchers expected that *non-uniform fields with strong gradients would be necessary to excite spin waves*. It was actually this non-uniformity that scientists initially relied on to explain the creation of spin waves. Kittel [29] pointed out that the *spins at the surface should be pinned by magnetic surface anisotropy*. This observation was incorporated in spin-wave analysis through the formulation of a new boundary condition at the surface of the sample. What is most important is that this condition enables the *excitation of spin waves* (in particular those of low orders) by *uniform microwave magnetic fields*. This is of primary practical importance in ferrite slabs and films. When both surfaces of the film are free (interfaced with air), only the odd ($k = 1, 3, 5, 7, \ldots$) spin eigenmodes are excited [8]. This is a direct consequence of the boundary conditions representing the spins pinned at both surfaces, which actually lead to a form of standing wave.

Moreover, the above approach inspired convenient experimental techniques, which provide direct evidence of the existence of spin waves. These are simply based on the detection of the resonance absorption of microwave energy when the ferrite sample is appropriately located inside a cavity and the DC biasing magnetic field is perpendicular to the film.

3.15   SPIN WAVES IN A FINITE SAMPLE

In the previous analysis, an infinitely extending sample, or at least a sample larger than the wavelength, was considered. In this case, the surface boundary conditions were not required to define the characteristic equation. Based on this approach, spin waves are approximated by plane waves. The above approach breaks down when the wavelength becomes smaller than the dimensions of the reciprocal sample. One may argue here that in this case we are already in the area of magnetostatic waves. This gray area may also be viewed as "finite sample spin waves" which are affected by the boundary conditions or the surface magnetic-charge density ($\hat{n} \cdot \bar{M}$). Their characteristic is that the demagnetization factors [N] are associated with these surface poles. According to Anderson and Suhl [31, 32] and [33], these spin waves can still be approximated by plane waves as above. Concerning the validity of this approach, Walker [10] stated that there will be cases under question where $\lambda \leq$ (sample dimensions)/10. Beyond this limit, the plane wave approximation is not valid. So, the field distribution should be obtained by solving the Laplace equation for the magnetostatic potential (Φ_m), imposing the appropriate boundary conditions on the sample surface.

For the plane-wave approach, the volume magnetic-charge density ($\nabla \cdot \bar{M}$) can be accounted for through the equivalent tensor of N-factors given in Eq. (3.45). These are a direct consequence of the dipolar field distribution, (3.39), which is valid only for the plane-wave approach. The surface charge density ($\hat{n} \cdot \bar{M}$) may be accounted for through the classical demagnetization factors. An example for certain ellipsoid of revolution samples is given in Eq. (3.10). In a first-order approximation, the most significant form of demagnetization is that imposed on the applied DC biasing field. That is in the same manner in which Eq. (3.13) is transformed into the Kittel's resonance of Eq. (3.14); that is, the field inside the sample along the DC bias z-direction reads as follows:

$$H_0 = H_{iz} = H_a - N_z M_s \tag{3.53a}$$

$$\omega_0 = \mu_0 \gamma H_0 = \mu_0 \gamma H_a - N_z \cdot \mu_0 \gamma M_s = \omega_a - N_z \cdot \omega_m, \tag{3.53b}$$

where H_a is the applied DC biasing field outside the sample, which reasonably resembles the DC magnetic field in the absence of the sample.

Substituting ω_0 in Eqs. (3.52), the spin-wave resonance reads as follows:

$$\omega_k = \left[\left(\omega_a - N_z \omega_m + \frac{Dk^2}{M_s} \omega_m \right) \cdot \left(\omega_a - N_z \omega_m + \frac{Dk^2}{M_s} \omega_m + \omega_m \sin^2 \theta_k \right) \right]^{1/2} \tag{3.54a}$$

or

$$\omega_k = \mu_0 \gamma \left[\left(H_a - N_z M_s + Dk^2 \right) \cdot \left(H_a - N_z M_s + Dk^2 + M_s \cdot \sin^2 \theta_k \right) \right]^{1/2}. \tag{3.54b}$$

Expressions (3.54) are the same as those given by [10, 31] and [33]. The only difference is that herein they are given in the SI system of units instead of the CGS system employed in the references.

Moving on to the effects of surface poles, the demagnetizing field of the alternating quantities (e.g., $\bar{m}$ and $\bar{H}_{ac}$) can be considered in a manner like that of (3.43). For this purpose, the demagnetization factors $N = (N_x, N_y, N_z)$ will be additionally considered. The N_z component will be accounted for through (3.53), while N_x and N_y will be added to the main diagonal of $[N]_{eq}$ given in (3.45). Remember that herein a ferrite biased to saturation in the z-direction is considered throughout the spin-wave section. In turn, the non-diagonal elements of $[N]_{eq}$ remain the same as those in (3.45) while its diagonal reads as follows:

$$\begin{aligned} N_{xx} &= N_x + \frac{Dk^2}{M_s} + \frac{k_x^2}{k^2} \\ N_{yy} &= N_y + \frac{Dk^2}{M_s} + \frac{k_y^2}{k^2}. \end{aligned} \tag{3.55}$$

The resulting magnetic susceptibility is again given by (3.46), but in conjunction with (3.55). Likewise, the characteristic equation results by setting $\Delta = 0$, which is given in (3.47). Exploiting (3.55), we have:

$$N_{xx} + N_{yy} = 2\frac{Dk^2}{M_s} + \sin^2 \theta_k + \left(N_x + N_y \right) \tag{3.56a}$$

$$N_\Delta = \frac{Dk^2}{M_s}\left(\frac{Dk^2}{M_s} + \sin^2\theta_k\right) + \frac{Dk^2}{M_s}\left(N_x + N_y\right) + N_x N_y + N_x \frac{k_y^2}{k^2} + N_y \frac{k_x^2}{k^2}, \quad (3.56b)$$

where θ_k is the angle between the wavenumber $\bar{k}$ and the z-axis. In turn, Eq. (3.47) reads as follows:

$$\Delta = 1 + \frac{Dk^2}{M_s}\mathcal{X}_{xx} + \left(\frac{Dk^2}{M_s} + \sin^2\theta_k + N_x + N_y\right)\cdot\left(\mathcal{X}_{xx} + \frac{Dk^2}{M_s}\mathcal{X}_\Delta\right)$$
$$+ \left(\frac{k_x^2}{k^2}N_y + \frac{k_y^2}{k^2}N_x + N_x N_y\right)\mathcal{X}_\Delta. \tag{3.57}$$

Recall that $k_x = k\sin\theta_k\cos\varphi_k$ and $k_y = k\sin\theta_k\sin\varphi_k$. Substituting $\mathcal{X}_{xx}$ from (3.44) and $\mathcal{X}_\Delta$ from (3.48), the characteristic equation $\Delta = 0$ yields a resonance frequency of:

$$\omega^2 = \omega_k^2 = \left(\omega_0 + \frac{Dk^2}{M_s}\omega_m\right)\cdot\left(\omega_0 + \frac{Dk^2}{M_s}\omega_m + \omega_m\sin^2\theta_k\right)$$
$$+ \left(N_x\cdot\sin^2\varphi_k + N_y\cdot\cos^2\varphi_k\right)\cdot\sin^2\theta_k\cdot\omega_m^2 + N_x N_y\cdot\omega_m^2. \tag{3.58a}$$

Many ferrite ellipsoidal specimens used in practical applications have a symmetry yielding $N_x = N_y = N_\perp$, which simplifies resonance as follows:

$$\omega_k^2 = \left(\omega_0 + \frac{Dk^2}{M_s}\omega_m\right)\left(\omega_0 + \frac{Dk^2}{M_s}\omega_m + \omega_m\sin^2\theta_k\right) + N_\perp\left(N_\perp + \sin^2\theta_k\right)\omega_m^2. \quad (3.58b)$$

The demagnetization factor in the z-direction can again be accounted for through (3.53) to yield:

$$\omega_k^2 = \left(\omega_a - N_z\omega_m + \frac{Dk^2}{M_s}\omega_m\right)\left(\omega_a - N_z\omega_m + \frac{Dk^2}{M_s}\omega_m + \omega_m\sin^2\theta_k\right)$$
$$+ N_\perp\left(N_\perp + \sin^2\theta_k\right)\omega_m^2 \tag{3.58c}$$

or

$$\omega_k = \mu_0\gamma\left[\left(H_a - N_z M_s + Dk^2\right)\left(H_a - N_z M_s + Dk^2 + M_s\sin^2\theta_k\right)\right.$$
$$\left. + N_\perp\left(N_\perp + \sin^2\theta_k\right)M_s^2\right]^{1/2}. \tag{3.58d}$$

Even though the latter expressions come as a logical follow-up of the above, we were not able to locate this analysis in the literature for ferromagnetic or ferrimagnetic specimens. However, for antiferromagnetic samples, the works of Loudon and Pincus [34], and Keffer and Kittel [35], account for all three demagnetization factors as above.

The next step in the analysis is to abandon plane-wave approximation and consider a more accurate distribution of the dipolar magnetic field, obeying the boundary conditions on the specimen surface. Fortunately, in this case, the wavelength is so large compared to the sample dimensions that propagation may be ignored. In other words, we are in the area of magnetostatic modes.

3.16   MAGNETOSTATIC WAVES

The term "magnetostatic waves" [30] refers to the low wavenumber ($k \to 0$) area of the spin-wave manifold. A qualitative difference between spin and magnetostatic waves is a wavenumber $k_{\min} = 2\pi/\lambda_{\max} \approx 10/(\text{sample dimensions})$. Therefore, for $k < k_{\min}$ or $\lambda > \lambda_{\max}$, one may seek magnetostatic waves. In turn, magnetostatic waves could be understood in a broad sense as being "*spin waves of finite samples.*" To be more specific, their wavelengths are so large that exchange interaction is negligible. In contrast, the dominant mechanism exciting magnetostatic waves is the dipole-dipole interaction. However, in this case, the sample dimensions are comparable to the wavelength and the field generated by the surface magnetic-charge (poles) density can effectively reach every point within the sample. Thus, both, volume ($\nabla \cdot \bar{M}$) and surface equivalent magnetic charge densities must be taken into account. Besides this, the phase velocity (V_{ph}) of magnetostatic modes is very small compared to the speed of light (c) as well as the speed of an electromagnetic wave ($c/\sqrt{\varepsilon_r \mu_r} = c/n$). In turn, for practically used samples propagation (phase difference from point to point) can be ignored and the so-called "magnetostatic approximation" can be employed in their analysis. Thus, magnetostatic modes are *shape dependent* due to surface poles ($\hat{n} \cdot \bar{M}$) but are *size independent* as propagation is ignored. Like electrostatics, the field quantities ($\bar{H}$, $\bar{M}$) are assumed to vary synchronously in time all over the sample. Sometimes, this definition is roughly considered. That happens because waves with inherently assumed propagation exist. More accurate is the use of magneto-quasi-static modes. In this case, the fields are in acceptable agreement with real-world measurements where some propagation exists.

To sum up, we can say that in the analysis, the magnetic surface-charge density ($\hat{n} \cdot \bar{M}$) should be taken into account ($\hat{n}$ is the unit outward normal at the sample-air interface). In other words, magnetostatic waves are coming from the magnetization discontinuity at the sample-air interface. Since the magnetic-field intensity inside the specimen (internal $\bar{H}_i$) is directly proportional to magnetization, the boundary condition is assigned to the discontinuity of the magnetic field. The same has already been done in Section 3.4, where the discontinuity was taken into account through the demagnetization factors [N]. The major difference in this case is that the volume magnetic charge is assumed to be equal to zero, or equivalently, magnetization is homogeneous, $\nabla \cdot \bar{M} = 0$. As the whole problem is about boundary conditions and, in particular, only shape dependent, significant research effort has been devoted to magnetostatic modes. Walker [9, 10] studied spheroid samples and established the procedure for magnetostatic-wave analysis. Also, Damon and Eshbasch [13], solved the problem for thin layers where the corresponding magnetostatic modes are often named after these authors.

3.17 SUSCEPTIBILITY AND CHARACTERISTIC EQUATION–UNIFORM MODE

The magnetization or LLG equation (3.23) will be solved following the same approach as in the previous cases, but the effective magnetic field ($\bar{H}^m$) will include both volume and surface-pole dipolar fields, while the exchange field will be ignored, reading as follows [30, 36, 37]:

$$\bar{H}^m = \bar{H}_e + \bar{H}_{dip} + \bar{H} + \bar{H}_{anis}. \tag{3.59}$$

However, it will be shown in the next paragraphs that this approach is only feasible for the uniform mode.

The simplest solution of the LLG equations refers to the *uniform mode*. In this mode, magnetization can be approximately considered uniform throughout the sample, resulting in a vanishing volume magnetic-charge density ($\nabla \cdot \bar{M} = 0$). In turn, the dipolar field is solely due to the surface magnetic charge ($\hat{n} \cdot \bar{m}$) or the discontinuity of magnetization at the sample-air interface ($\hat{n}$ is the outward unit normal). In turn, the dipolar field $\left(\bar{H}_{dip}^0 \right)$ is given by (3.34) through the second part of (3.37), or equivalently [24], by

$$\bar{H}_{dip}^0 = \frac{1}{4\pi} \oiint_s \frac{\left[\hat{n} \left(\bar{r}' \right) \cdot \bar{M} \left(\bar{r}' \right) \right] \left(\bar{r} - \bar{r}' \right)}{\left| \bar{r} - \bar{r}' \right|^3} ds'. \tag{3.60a}$$

In the limiting case of a film or plate with longitudinal dimensions much larger than its thickness, the above dipolar field can be approximated as follows [30]:

$$\text{Film:} \quad \bar{H}_{dip}^0 \approx -\hat{n} \left(\hat{n} \cdot \bar{M}_s \right), \tag{3.60b}$$

where $\hat{n}$ is the outward unit vector normal to the film or plate.

However, we should remember Kittel's classical approach [2], in which this dipolar demagnetizing field (or shape anisotropy field) is accounted for through the demagnetization factors $[N]$

$$\bar{H}_{dip}^0 = -[N]\bar{M}, \tag{3.60c}$$

where $[N]$ is in general a tensor.

Kittel's classical approach only considers diagonal elements. In this case, magnetic susceptibility has already been given in Eqs. (3.9) through (3.11), while the corresponding uniform mode resonance frequency appears in (3.14) through (3.16). McDonald [36] employed a tensor $[N]$, but he states that Kittel's formula is a good approximation even when the non-diagonal elements are non-zero. He provided that the DC biasing field $\bar{H}_{DC}$ is large enough to align $\bar{M}$ parallel to its direction.

The demagnetization factors, for certain useful shapes (sphere, rod, disk or plate) approximated by suitable ellipsoids of revolution, are given in Eq. (3.8). According to [17], for prolate or oblate spheroids we have:

$$N_t = \frac{ab}{2c^2} \cdot \left[\frac{b}{a} - \frac{a}{c} \sinh^{-1} \left(\frac{c}{a} \right) \right] \quad \text{for } a < b \qquad (3.61a)$$

$$N_t = \frac{ab}{2c^2} \cdot \left[\frac{a}{c} \sinh^{-1} \left(\frac{c}{a} \right) - \frac{b}{a} \right] \quad \text{for } a > b \qquad (3.61b)$$

$$N_Z = 1 - 2N_t, \qquad (3.61c)$$

where a and b are the relevant semi-axes of their elliptical cross-section and $c = \left[|a^2 - b^2| \right]^{1/2}$.

The next step is to account for the non-uniformity of magnetization $\bar{M}$ or the volume magnetic charge ($\nabla \cdot \bar{M} \neq 0$), where the dipolar magnetic field becomes non-uniform. This is actually a far more complicated situation, where a Laplace equation for the magnetostatic potential must be solved by imposing the appropriate continuity conditions on the specimen's surface. Equivalently, one could include Eq. (3.36) in a numerical simulation model. Herein, only certain canonical practical shapes (disk, film, sphere, and cylinder), for which classical analytical solutions are available in the literature, will be considered. The primary aim is to identify the characteristics of magnetostatic modes in order to develop the ability to exploit them in practical microwave applications.

3.18   THE MAGNETOSTATIC EQUATION OF A UNIFORMLY BIASED SPECIMEN

The electromagnetic field of a uniformly biased specimen is effectively magnetostatic. Therefore, the resonances observed are just the free modes of oscillation of an array of magnetic dipoles in a uniform magnetic field. Focusing on the magnetostatic modes of a spheroid, we must say that these belong to its normal modes, called "Walker modes" after the researcher who originally studied them. Note that a spheroid is a sphere with one of its dimensions altered, either squashed toward a disk shape (oblate) or extended like a cigar (prolate). The analysis below will follow Walker only in a general form. Avoiding numerical details, we will clarify the main characteristics of magnetostatic modes.

At this point, it is necessary to recast the main equations in order to adapt them to the magnetostatic regime and to Walker's analysis. Concerning the magnetic field, it contains a DC term internal to the specimen, $\bar{H}_i = \bar{H}_{DC}$, and a time-harmonic alternating term ($\bar{h}e^{j\omega t}$):

$$\bar{H} = \bar{H}_i + \bar{h}e^{j\omega t} = H_i \hat{z} + \bar{h}e^{j\omega t}. \qquad (3.62)$$

A $\hat{z}$-biased ferrite magnetized to saturation has already been assumed above while the alternating component may also have a z-component, but this is usually negligible under the small-signal approximation, $|\bar{h}| \leq |\bar{H}_i|$. Likewise, magnetization is divided into a static component, the saturation magnetization along the z-axis, $\bar{M}_{DC} = \bar{M}_s = M_s \hat{z}$, and a time-harmonic component $\bar{m}e^{j\omega t}$:

$$\bar{M} = \bar{M}_s + \bar{m}e^{j\omega t} = M_s \hat{z} + \bar{m}e^{j\omega t}. \tag{3.63}$$

Again, under the small-signal approximation, the z-component of $\bar{m}$ is $m_z << M_s$. Thus, $\bar{m}$ can be assumed transverse to the z-axis.

A very important simplification results from the practical observation that the DC components $(\bar{H}_i, \bar{M}_s)$ can be assumed to be uniform throughout the specimen. In turn, the related volume magnetic-charge density vanishes $(\nabla \cdot \bar{M}_s = 0)$. Therefore, the internal magnetic field is affected only by the surface magnetic charge, which can be accounted for through the demagnetization factors $[N]$ as $\bar{H}_i = \bar{H}_0 - [N]\bar{M}$. However, since the DC bias is assumed along the $\hat{z}$-axis, we have

$$\bar{H}_i = \bar{H}_0 - N_z M_s \hat{z}. \tag{3.64}$$

The external microwave field will be considered through the boundary conditions at the specimen's surface to be imposed on Maxwell equations for $(\bar{m}, \bar{h})$. Likewise, the dipolar field contribution is inherently included within these equations. The only remaining term is the anisotropy field $\bar{H}_{anis}$. As far as crystalline samples are concerned and, in particular, single crystals magnetized along the "easy" or "hard" direction, the anisotropy field has already been shown to be inversely proportional to M_s. That is,

$$\bar{H}_{anis} = \frac{2k}{M_S^2} M_z \hat{z} \approx \frac{2k}{M_S} \hat{z}. \tag{3.65}$$

Consequently, assuming the above, $\bar{H}_{anis}$ acts exactly as an additional term of the DC magnetic field and just like in the previous cases, it can be absorbed in $\bar{H}_0$. That is,

$$\bar{H}_0 \leftarrow \bar{H}_0 + \bar{H}_{anis} \quad \text{or} \quad \bar{H}_i = \bar{H}_0 + \bar{H}_{anis} = (H_0 + H_{anis})\,\hat{z}. \tag{3.66}$$

In contrast to the above assumed DC behavior, the alternating quantities $(\bar{m}, \bar{h})$ are non-uniform throughout the specimen, giving rise to volume $(\nabla \cdot \bar{m} \neq 0)$ as well as to surface $(\hat{n} \cdot \bar{m} \neq 0)$ magnetic-charge densities. The worst is that the field distribution depends on the specimen's shape, requiring the solution of a magnetostatic problem subject to the boundary conditions on the specimen's surface (sample-air interface). Sometimes, even plane-wave approximation may be employed to get some qualitative results, but these are very rough and far from accurate. However, it has proved very useful to seek the expansion of the magnetostatic solution in the superposition of a large number of plane waves. In addition, keep in mind that the magnetostatic solution is expected to be size independent since it ignores propagation phenomena. Returning to Walker's modes, the alternating components $(\bar{m}, \bar{h})$ should satisfy Maxwell's equations in their reduced magnetostatic form. We define the corresponding alternating flux density $\bar{b}$ as

$$\bar{b} = \mu_0 \left(\bar{h} + \bar{m} \right). \tag{3.67}$$

Once again, following the reasoning of magnetostatics, Maxwell's equations within the specimen read as follows:

$$\nabla \cdot \bar{b} = 0 \quad \leftrightarrow \quad \nabla \cdot \bar{h} = -\nabla \cdot \bar{m} = \rho_m \tag{3.68a}$$

$$\nabla \times \bar{h} = 0 \quad \leftrightarrow \quad \bar{h} = -\nabla \psi \tag{3.68b}$$

and $\qquad \nabla^2 \psi (\bar{r}) = -\rho_m (\bar{r}) \tag{3.68c}$

or $\qquad \nabla^2 \psi (\bar{r}) - \nabla \cdot \bar{m} (\bar{r}) = 0. \tag{3.68d}$

Besides the above, the magnetization or LLG equation should be satisfied as well. We have

$$\frac{d \bar{M}}{dt} = -\gamma \mu_0 \bar{M} \times \bar{H}. \tag{3.69}$$

Let us consider (3.69) through (3.62) and (3.63) under small-signal approximation. It is obvious that $\bar{m}$ can be related to $\bar{h}$ through the same susceptibility tensor by just substituting $H_0 \rightarrow H_i$ or $\omega_0 = \mu_0 \gamma H_0 \rightarrow \omega_i = \mu_0 \gamma H_i$. Thus, for the present case, (2.17) reads

$$\bar{m} = [\mathcal{X}] \bar{h} = \begin{bmatrix} \mathcal{X}_{xx} & \mathcal{X}_{xy} & 0 \\ \mathcal{X}_{yx} & \mathcal{X}_{yy} & 0 \\ 0 & 0 & 0 \end{bmatrix} \bar{h} \tag{3.70}$$

and

$$\mathcal{X}_{xx} = \mathcal{X}_{yy} = \mathcal{X} = \frac{\omega_i \omega_m}{\omega_i^2 - \omega^2} = \frac{\Omega_H}{\Omega_H^2 - \Omega^2} \tag{3.71a}$$

$$\mathcal{X}_{xy} = \mathcal{X}_{yx} = jk_1 = j \frac{\omega \omega_m}{\omega_i^2 - \omega^2} = j \frac{\Omega}{\Omega_H^2 - \Omega^2} \tag{3.71b}$$

where

where $\qquad \Omega = \omega / \omega_m \quad \text{and} \quad \Omega_H = \omega_i / \omega_m. \tag{3.71c}$

Likewise, the flux density vector $\bar{b}$ is related to $\bar{h}$ through the permeability tensor:

$$\bar{b} = \mu_0 \{[I] + [\mathcal{X}]\} \bar{h} = [\mu] \bar{h} \tag{3.72}$$

and

$$[\mu] = \mu_0 \begin{bmatrix} 1 + \mathcal{X} & jk_1 & 0 \\ -jk_1 & 1 + \mathcal{X} & 0 \\ 0 & 0 & 1 \end{bmatrix}. \tag{3.73}$$

Even though the above seem rather complicated, the actual situation is much simpler since only Eqs. (3.72), (3.68a), and (3.68b) are needed to yield:

$$\text{Ferrite Region:}\quad \nabla \cdot \left([\mu]\bar{h}\right) = 0 \;\leftrightarrow\; \nabla \cdot \left([\mu]\nabla\Psi\right) = 0. \tag{3.74}$$

This is the equation governing the magnetostatic potential inside the ferrite specimen. However, for the air region surrounding the specimen, magnetization is zero, or permeability is scalar and equal to μ_0, or $\bar{b} = \mu_0\bar{h}$ and (3.74) is reduced to:

$$\text{Air Region:}\quad \nabla^2\psi = 0. \tag{3.75}$$

What is then needed is to solve the generalized Laplace equation (3.74) inside the specimen and the ordinary one (3.75), outside the specimen, and impose the boundary conditions on its surface. For an analytical solution, the unknown constants will be estimated through the continuity conditions. Let the superscripts i and e denote the inside and outside external regions respectively. Thus, we have

$$\text{Ferrite Region:}\quad \nabla \cdot \left([\mu]\nabla\Psi^i\right) = 0 \tag{3.76a}$$

$$\text{Air Region:}\quad \nabla^2\Psi^e = 0. \tag{3.76b}$$

On the specimen's surface with a normal unit vector $\hat{n}$, we impose the following conditions:

1. Continuity of tangential intensity components $\hat{n} \times \bar{h}$.

2. Continuity of normal flux density b_n or $\hat{n} \cdot \bar{b}$.

3. The magnetic potential ψ^e should be bounded (vanish) at infinity $\lim \psi^e(r \to \infty) \to 0$.

Equations (3.76) may be expanded in Cartesian coordinates in order to obtain their general solutions. It is worth noting that the permeability elements are assumed to be independent of position since $\mathcal{X}$ and k_1 depend on $\omega_i = \mu_0\gamma H_i$ and $\omega_m = \mu_0\gamma M_s$, while H_i and M_s are assumed to be uniform throughout the specimen. In turn, (3.76a) reads as follows:

$$(1 + \mathcal{X}) \cdot \left(\frac{\partial^2\psi^i}{\partial x^2} + \frac{\partial^2\psi^i}{\partial y^2}\right) + \frac{\partial^2\psi^i}{\partial z^2} = 0. \tag{3.77}$$

This is actually Walker's equation for magnetostatic modes of uniformly biased (or homogeneous) ferrite specimens.

3.19 MAGNETOSTATIC MODES IN AN INFINITE MEDIUM

The study of magnetostatic modes in an infinity medium does not reflect any useful practical applications. However, it serves two significant purposes. First, it clarifies some of their characteristics and, second, it can be incorporated into the analysis of practical finite specimens, where the magnetic potential is expanded into an infinite sum of plane waves (just like a Fourier series).

Let us now consider a uniform plane wave propagating in an arbitrary direction within a uniformly $\hat{z}$-biased infinite ferrite medium. The general solutions of (3.77) will take the following form:

$$\psi = \psi_0 e^{-j\bar{k}\cdot\bar{r}} \tag{3.78}$$

$\bar{k} = k_x\hat{x} + k_y\hat{y} + k_z\hat{z}$ and $\partial/\partial x = -jk_x$, $\partial/\partial y = -jk_y$ and $\partial/\partial z = -jk_z$. In turn, Eq. (3.77) reduces to:

$$(1 + \mathcal{X})\left(k_x^2 + k_y^2\right) + k_z^2 = 0. \tag{3.79}$$

Let vector $\bar{k}$ point to a direction forming an angle θ with respect to the biasing $\hat{z}$-direction. Then, $k_x^2 + k_y^2 = k^2 \sin^2 \theta$ and $k_z^2 = k^2 \cos^2 \theta$. Equation (3.79) becomes

$$(1 + \mathcal{X})k^2 \sin^2 \theta + k^2 \cos^2 \theta = 0 \quad \leftrightarrow \quad \mathcal{X}\sin^2 \theta + 1 = 0. \tag{3.80}$$

One may substitute $\mathcal{X}$ from (3.71) and solve either for the angle θ in terms of frequency ω or vice versa. In the latter case, we have

$$\omega_i \omega_m \sin^2 \theta = -\omega_i^2 + \omega^2 \tag{3.81a}$$

$$\text{or} \quad \omega = \sqrt{\omega_i \left(\omega_i + \omega_m \sin^2 \theta\right)}. \tag{3.81b}$$

Equation (3.81b) can be read as "a magnetostatic mode occurring at a specific frequency ω and propagating at a specific angle-θ and vice versa." This is a well-known phenomenon in waveguides, where each mode can be considered as a plane wave propagating in a zigzag manner after successive reflections at the guide walls, but always at a specific constant angle with respect to the propagation axis. Equation (3.81b) could then be interpreted as representing a "*self-guiding*" property of magnetostatic modes in a uniformly magnetized ferrite.

Note, also the resemblance of (3.81b) to the spin-wave resonance equation. In fact, the spin-wave resonance equation reduces to (3.81b) when either exchange interaction is ignored (as herein), or when the wavenumber tends to zero $k \rightarrow 0$. More importantly, the magnetostatic mode frequency (ω) appears independent of wavenumber k. This means that multiple degenerate modes exist at the same frequency and all wavelengths have the same energy. This degeneracy is removed when exchange interaction is taken into account or when the boundaries of a practical finite specimen are considered.

3.20   MAGNETOSTATIC MANIFOLD

A common straightforward approach employed for the derivation of an analytical solution of the magnetostatic equation relies on the use of the separation of variables. This also yields the conditions under which magnetostatic modes exist or defines their manifold. For this purpose, let the magnetic potential inside (ψ^i) and outside (ψ^e) the specimen be separated as follows:

$$\psi^i = \psi^i(x, y, z) = f_i(x)g_i(y)h_i(z) \tag{3.82a}$$

$$\psi^e = \psi^e(x, y, z) = f_e(x)g_e(y)h_e(z). \tag{3.82b}$$

The primed symbols denote derivatives (e.g., $f''(x) = d^2 f/dx^2$). We substitute (3.82) into (3.76) and get

$$(1 + \mathcal{X}) \cdot \left(\frac{f_i''}{f_i} + \frac{g_i''}{g_i} \right) + \frac{h_i''}{h_i} = 0 \tag{3.83a}$$

$$\frac{f_e''}{f_e} + \frac{g_e''}{g_e} + \frac{h_e''}{h_e} = 0. \tag{3.83b}$$

As is often the case when exponential, sinusoidal or hyperbolic solutions are sought, we may denote

$$f''/f = -k_x^2, \quad g''/g = -k_y^2, \quad h''/h = -k_z^2. \tag{3.84}$$

Substituting expression (3.84) to (3.83a) and (3.83b), we have

$$(1 + \mathcal{X}) \left(k_{xi}^2 + k_{yi}^2 \right) + k_{zi}^2 = 0 \tag{3.85a}$$

$$k_{xe}^2 + k_{ye}^2 + k_{ze}^2 = 0. \tag{3.85b}$$

Without loss of generality, consider a partial solution of $f''/f = -k_x^2$ in its exponential form $e^{-jk_x x}$, which corresponds to a positive k_x^2 and represents a uniform wave propagating in the positive x-direction. In contrast, when k_x^2 is negative, this solution takes the form of $e^{-\alpha_x x}$, where $k_x^2 = (j\alpha_x)^2 = -\alpha_x^2 < 0$ and $\alpha_x > 0$, which then represents a wave exponentially decaying in the positive x-direction.

In view of the above, let us consider the possible solutions of (3.76) through (3.85). From (3.85b) it is obvious that is impossible to enforce k_{xe}^2, k_{ye}^2, k_{ze}^2 to be either all positive or all negative. At least one of them should be negative and the other two positive, or vice versa. This means that in the air region surrounding the ferrite specimen, there will be an exponential decay in at least one direction and propagation along the two other directions, or vice versa. In the most common cases, the exponential attenuation occurs in the direction normal to the ferrite specimen surface while the wave propagates along its surface. These modes are classified as surface waves when ordinary materials are considered. However, this does not apply to magnetostatic modes since all possible modes retain this property. In contrast, in the interior of the ferrite specimen, there is an additional degree of freedom. This comes from the term $(1 + \mathcal{X})$, which depends on the value of susceptibility $\mathcal{X}$. To be more exact, the degree of freedom depends on the sign of $(1 + \mathcal{X})$ that controls the type of wave. Thus, when $(1 + \mathcal{X}) > 0$, the same principles as for the air region apply; in other words, it should be an exponential decay in at least one direction, and the corresponding k_w^2 $(w = x, y, z)$ will be negative. When this decay direction is normal to the specimen's surface, the wave energy will be confined to a very small distance from the surface and the wave will propagate around the specimen's surface, enabling particularly useful applications. This is exactly the type of mode which is classified as "surface magnetostatic mode."

Besides the above, it is possible for $(1 + \mathcal{X})$ to be negative, which enables all three k_{xi}^2, k_{yi}^2, k_{zi}^2 to be positive, allowing propagation in all directions and their field to exist all over the specimen. This type of solution is usually called "volume magnetostatic waves." It is interesting to define the frequency range of these modes by using (3.71) for $\mathcal{X}$ as follows:

$$1 + \mathcal{X} < 0 \quad \leftrightarrow \quad \mathcal{X} = \frac{\omega_i \omega_m}{\omega_i^2 - \omega^2} < -1. \tag{3.86}$$

This inequality may have two mathematical solutions:

$$\omega > \omega_i \quad \leftrightarrow \quad \omega_i^2 - \omega^2 < 0 \quad \leftrightarrow \quad \omega < \sqrt{\omega_i (\omega_i + \omega_m)} \tag{3.87a}$$

$$\omega < \omega_i \quad \leftrightarrow \quad \omega_i^2 - \omega^2 > 0 \quad \leftrightarrow \quad \omega > \sqrt{\omega_i (\omega_i + \omega_m)}. \tag{3.87b}$$

Usually, $\omega_i = \mu_0 \gamma H_i$ (where $H_i = H_0 - N_z M_s + H_{anis}$) is smaller than $\omega_m = \mu_0 \gamma M_s$ and the first solution applies to yield:

$$\text{Volume magnetostatic wave manifold:} \quad \omega_i \ < \ \omega \ < \ \sqrt{\omega_i (\omega_i + \omega_m)}. \tag{3.88}$$

At this point, we may proceed to the study of magnetostatic modes of an infinitely extending thin slab or film. The reason is the more convenient rectangular coordinates employed, while on the other hand, thin slabs and thin films are of primary importance in the trend toward miniaturization and integration of microwave circuits.

3.21   MAGNETOSTATIC MODES OF AN INFINITELY EXTENDING THIN SLAB FILM

Let us begin with the simple case of transverse magnetization of infinitely extending thin slabs. Due to their symmetry, this is the easiest case.

The coordinate system with respect to the thin ferrite slab is shown in Figure 3.11. In accordance with Walker's theory and notation, the DC biasing field is assumed along the z-axis, which for the present case is oriented normally to the slab infinitely extending along the x- and y-axes.

Concerning the internal DC magnetic field, it is $H_i = H_0 - M_s$ since the slab demagnetization factors from Eq. (3.8c) are $N_x = N_y = 0$, $N_z = 1$. As the film is infinitely extending in the x- and y-directions, the corresponding dependence of the field inside the slab can

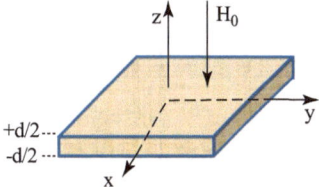

Figure 3.11: An infinitely extending thin ferrite slab normally biased to saturation.

just be a propagating term such as $e^{-jk_x x} \cdot e^{-jk_y y} = e^{-j\bar{k}_t \cdot \bar{\rho}}$, where $\bar{k}_t = k_x \hat{x} + k_y \hat{y}$ while $\bar{\rho} = x\hat{x} + y\hat{y}$. In contrast, the slab-air interfaces at $z = \pm d/2$ impose a standing-wave solution. In turn, a general solution to the magnetic potential could be:

$$\psi^i = \left(A_z e^{-jk_{zi}\cdot z} + B_z e^{jk_{zi}\cdot z} \right) e^{-j\bar{k}_t \cdot \bar{\rho}}. \tag{3.89}$$

Condition (3.85a) reads as follows:

$$(1 + \mathcal{X})k_t^2 + k_{zi}^2 = 0. \tag{3.90}$$

For the general solution for ψ^e in the semi-infinite air (or any other dielectric) region, above ($z > +d/2$) or below ($z < -d/2$) the slab, we may take a step forward. This means that the continuity conditions at $z = \pm d/2$ require the same dependence in the x- and y-directions because they should be valid at any arbitrary (x, y) point on these surfaces. Moreover, in the z-direction, the potential should vanish at infinity, while the analysis in the previous section requires k_{ze}^2 to be negative for (3.85b) to be fulfilled. Therefore, let $k_{ze}^2 = (j\alpha)^2 = -\alpha^2 < 0$ and $\alpha > 0$ to get

$$\psi^e = \begin{cases} \psi_{0e}\, e^{-j\bar{k}_t \cdot \bar{\rho}}\, e^{-\alpha(z-d/2)} & \text{for } z \geq d/2 \\ \psi'_{0e}\, e^{-j\bar{k}_t \cdot \bar{\rho}}\, e^{+\alpha(z+d/2)} & \text{for } z \leq -d/2. \end{cases} \tag{3.91}$$

Futhermore, condition (3.85b) requires

$$k_t^2 - \alpha^2 = 0 \quad \leftrightarrow \quad k_t^2 = \alpha^2 \quad \leftrightarrow \quad \alpha = k_t. \tag{3.92}$$

Note that the positive sign is taken since α must be positive for the solution to be bounded at infinity.

Moreover, the above expressions should satisfy the boundary conditions at the ferrite air interface $z = \pm d/2$, which can be realized through (3.68b) and (3.72) requiring $\bar{h} = -\nabla\Psi$ and $\bar{b} = [\mu]\bar{h}$. Thus, from the continuity of the tangential magnetic field h_x, h_y at $z = \pm d/2$, we have

$$h_x = -\frac{\partial \psi}{\partial x} \quad \leftrightarrow \quad \frac{\partial \psi^i}{\partial x}\bigg|_{z=\pm d/2} = \frac{\partial \psi^e}{\partial x}\bigg|_{z=\pm d/2} \tag{3.93a}$$

$$h_y = -\frac{\partial \psi}{\partial y} \quad \leftrightarrow \quad \frac{\partial \psi^i}{\partial y}\bigg|_{z=\pm d/2} = \frac{\partial \psi^e}{\partial y}\bigg|_{z=\pm d/2}. \tag{3.93b}$$

In addition, from the continuity of the normal flux density $b_n = b_z$, we have

$$b_z = \mu_0 \cdot h_z = -\mu_0 \frac{\partial \psi}{\partial z} \quad \leftrightarrow \quad \frac{\partial \psi^i}{\partial z}\bigg|_{z=\pm d/2} = \frac{\partial \psi^e}{\partial z}\bigg|_{z=\pm d/2}. \tag{3.94}$$

In view of the symmetry at $z = \pm d/2$, also denoted by (3.94), one may intuitively choose a symmetric or cosinus solution for ψ^i, which also asks for $\psi_{0e} = \psi'_{0e}$ in (3.91). However, an antisymmetric or sinusoidal solution for ψ^i can be taken, too, which also asks for $\psi'_{0e} = -\psi_{0e}$ in (3.91). For the symmetric solution, let $A_z = B_z = \psi_{0i}/2$ in (3.89), which then reads as follows:

$$\psi^i = \psi_{0i} \cos(k_{zi}z)e^{-j\bar{k}_t \cdot \bar{\rho}}. \tag{3.95}$$

This greatly simplifies the algebra involved in the boundary conditions. The symmetric and anti-symmetric solutions could also be extracted mathematically by preserving (3.89) as is, but the algebra involved would be more complicated. In turn, (3.94) at either $z = d/2$ or $z = -d/2$ asks for

$$\psi_{0i}k_{zi} \cdot \sin(k_{zi} \cdot d/2) = \alpha \cdot \psi_{0e}. \tag{3.96}$$

While both conditions (3.93a) and (3.93b) lead to the same expression

$$\psi_{0i} \cos(k_{zi} \cdot d/2) = \psi_{0e} \tag{3.97}$$

by dividing (3.96) by (3.97) the unknown amplitudes are removed to yield

$$\tan(k_{zi} \cdot d/2) = \alpha/k_{zi}. \tag{3.98}$$

Substituting (3.92) into (3.90), we get

$$k_{zi}^2 = -(1 + \mathcal{X}) \cdot \alpha^2. \tag{3.99}$$

The sign of $(1 + \mathcal{X})$ in Eq. (3.99) is of primary importance and explicitly defines the type of magnetostatic wave. When a wave propagating in the z-direction is sought, k_{zi}^2 should be positive, and this is only possible when $(1 + \mathcal{X})$ is negative. The corresponding frequency range is defined by (3.87a) as follows:

$$\omega_i < \omega < [\omega_i(\omega_i + \omega_m)]^{1/2}. \tag{3.100}$$

It is obvious that in this case the wave energy will spread all over the ferrite slab, and for this reason, the solutions are called "magnetostatic volume waves (MSVW)." Moreover, propagation occurs in the three dimensions toward the positive direction (note that in the z-direction, a standing-wave pattern occurs due to the partial "reflections" at $z = \pm d/2$). In addition, it will be proved next that the phase and group velocities are both positive, denoting energy flow in the direction of propagation. For this reason, they are specifically called "Forward-MSVW, or MSFVW."

In contrast to the above case, when $(1 + \mathcal{X})$ is positive, k_{zi}^2 becomes negative, yielding a hyperbolic dependence in the z-direction, where the wave energy is concentrated around the surface both in the air and the ferrite medium. This case will be studied separately, but the important expected characteristic is that the wave propagates only in one direction (unidirectional) and not in the opposite one. Important applications termed "edge mode" are based on this phenomenon.

Returning to the MSFVW choice where $(1 + \mathcal{X}) < 0$, one may solve (3.99) in terms of k_{zi} as

$$k_{zi} = \alpha \cdot \sqrt{-(1 + \mathcal{X})}. \tag{3.101}$$

This can then be incorporated into (3.98) to yield the following characteristic equation of MSFVW modes:

$$\tan \left\{ \frac{\alpha d}{2} \sqrt{-(1 + \mathcal{X})} \right\} = \frac{1}{\sqrt{-(1 + \mathcal{X})}}. \tag{3.102}$$

The above procedure can be repeated for the anti-symmetric solution by letting $B_z = -A_z = \psi_{0i}/2j$ and $\psi'_{0e} = -\psi_{0e}$ in (3.91). Thus, we have:

$$\text{Anti-symmetric:} \quad \psi^i = \psi_{0i} \sin(k_{zi}z) e^{-j\bar{k}_t \cdot \bar{\rho}}. \tag{3.103}$$

In turn, the boundary conditions (3.93) and (3.94) yield

$$\psi_{0i} \sin(k_{zi}d/2) = \psi_{0e} \tag{3.104a}$$
$$\psi_{0i} k_{zi} \cos(k_{zi}d/2) = -\alpha \psi_{0e}. \tag{3.104b}$$

By dividing them, the characteristic equation for the anti-symmetric case is:

$$\cot(k_{zi}d/2) = -\alpha/k_{zi}. \tag{3.105}$$

Using the addition formulas for $\sin(z - n\pi/2)$ and $\cos(z - n\pi/2)$, one can easily prove that:

$$\tan(z - n\pi/2) = \begin{cases} -\cot(z) & \text{for } n = 1, 3, 5, \dots \\ \tan(z) & \text{for } n = 0, 2, 4, \dots. \end{cases} \tag{3.106}$$

In view of (3.106), the symmetric (3.98) and anti-symmetric (3.105) can be combined into a single characteristic equation as follows:

$$\text{MSFVW waves:} \quad \tan\left(k_{zi}\frac{d}{2} - \frac{n\pi}{2}\right) = \frac{\alpha}{k_{zi}} \tag{3.107a}$$

or

$$\text{MSFVW waves:} \quad \tan\left(\frac{\alpha d}{2}\sqrt{-(1+\mathcal{X})'} - \frac{n\pi}{2}\right) = \frac{1}{\sqrt{-(1+\mathcal{X})'}}, \tag{3.107b}$$

where $\mathcal{X}$ is given by (3.71). For $n =$ even, we have symmetric modes, whereas for $n =$ odd, we have anti-symmetric modes.

An approximate solution of (3.107b) for the lowest-order symmetric mode ($n = 0$) is:

$$\omega^2 = \omega_i\left\{\omega_i + \omega_m \cdot \left(1 - \frac{1 - e^{-\alpha d}}{\alpha d}\right)\right\}. \tag{3.108}$$

A graphic solution of the two characteristic equations in a manner similar to that classically used for the surface waves of a grounded dielectric slab (e.g., Pozar [3]) would greatly clarify the multiplicity of the excited modes and their characteristics as well. For this purpose, let us recast the characteristic equations in an appropriate form:

$$\text{Symmetric MSFVW:} \quad \left(k_{zi}\frac{d}{2}\right) \cdot \tan\left(k_{zi}\frac{d}{2}\right) = \left(\alpha \cdot \frac{d}{2}\right) \tag{3.109a}$$

$$\left(k_{zi}\frac{d}{2}\right) / \sqrt{-(1+\mathcal{X})} = \left(\alpha \cdot \frac{d}{2}\right) \tag{3.109b}$$

$$\text{Anti-symmetric MSFVW:} \quad -\left(k_{zi}\frac{d}{2}\right) \cdot \cot\left(k_{zi}\frac{d}{2}\right) = \left(\alpha \cdot \frac{d}{2}\right) \tag{3.110a}$$

$$\left(k_{zi}\frac{d}{2}\right) / \sqrt{-(1+\mathcal{X})} = \left(\alpha \cdot \frac{d}{2}\right). \tag{3.110b}$$

Each of the pairs of Eqs. (3.109) or (3.110) can be plotted on a diagram with axes $(\alpha d/2)$ vs. $(k_{zi}d/2)$. Expressions (3.109a) and (3.110a) contain the functions $(x \tan x)$ and $-(x \cot x)$, which present multiple branches. Their multiple points of intersection with the identical straight lines (3.109b) and (3.110b), whose slope is $1/\sqrt{-(1+\mathcal{X})}$, constitute the valid wavenumber pairs (α, k_{zi}) for each magnetostatic mode. Also keep in mind that only positive $\alpha = k_t$ represents valid solutions. These graphic representations are shown in Figures 3.12 and 3.13.

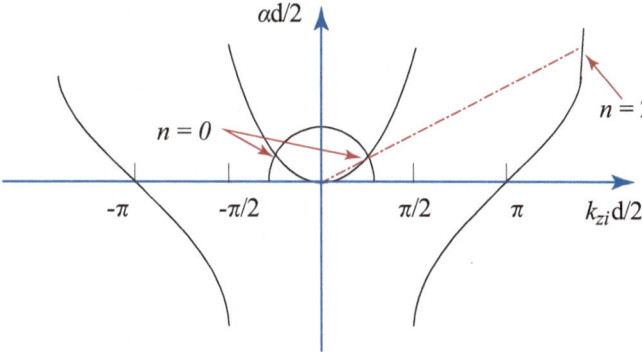

Figure 3.12: A graphic solution of the transcendental equation of symmetric MSFVW.

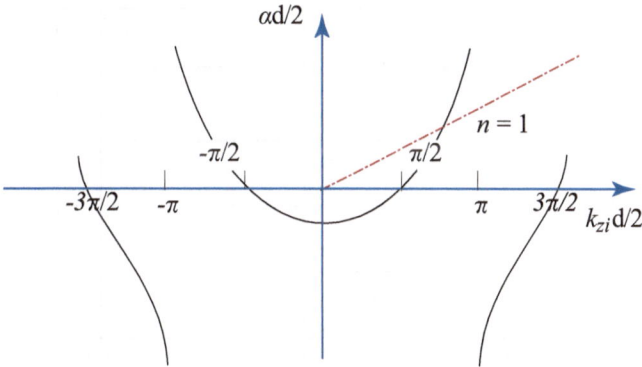

Figure 3.13: A graphic solution of the transcendental equation of anti-symmetric MSFVW.

3.21.1   PHASE AND GROUP VELOCITIES OF MSFVW

The term of propagation either in the ferrite film or in the air above and below is $\psi \propto e^{-jk_t\rho}e^{j\omega t}$.

In turn, it is very easy to define the phase (V_p) and group (V_g) velocities in the classical way as follows:

$$V_p = \frac{\omega}{k_t} \quad \text{and} \quad V_g = \frac{\partial \omega}{\partial k_t}. \tag{3.111}$$

In all the above expressions, $k_t = \alpha$. Also, remember that the important decision here is whether to classify these modes as forward or backward. This depends on the sign of the group velocity. Thus, positive means forward and negative means backward. This can also be seen clearly from the slope (V_g) of the dispersion curves $\omega - k_t$. For a mathematical expression, we may modify (3.107b) with respect to $\alpha = k_t$ and ω. This yields to

$$\frac{1}{V_g} = \frac{\mathcal{X} k_1}{(1 + \mathcal{X}) \cdot \omega_m d} \cdot \left[\frac{2}{\mathcal{X}} - k_t d \right]. \tag{3.112a}$$

For very thin slabs or films ($k_t d \ll 1$) using the definitions of k_1 and $\mathcal{X}$ from Eqs. (3.71), we have:

$$\left. \frac{1}{V_g} \right|_{k_{td} \to 0} \approx \frac{4}{\omega_m d}. \tag{3.112b}$$

The inverse of group velocity is plotted in Figure 3.14, where one can clearly see that $1/V_g$, and, in turn, V_g is always a positive quantity. Consequently, these are forward waves, where energy travels in the same direction as that of wave propagation.

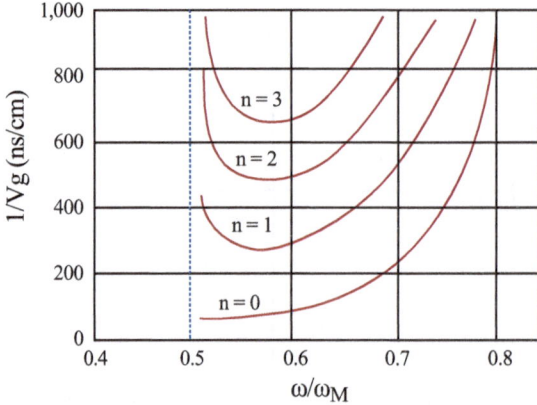

Figure 3.14: The inverse group velocity of MSFVW [12].

It would be useful to give an indicative figure. A YIG film with thickness $d = 10$ μm has a group velocity of the order of $V_g (n = 0) \approx 10^7$ m/sec $= c/30$ [17]; that is, this is a wave with a group velocity 30 times lower than that of an electromagnetic wave propagating in free space.

To conclude this section, it would be useful to observe how the wave direction of propagation is related to slab thickness. Bearing in mind that $\alpha = k_t$ and $\tan(x) \approx x$ when $x \ll 1$ and taking the limit of (3.107a) and (3.109a) for electrically thin slabs ($kd \ll 1$), we have:

Symmetric MSFVW for $kd \ll 1$: $(k_{zi} d/2)^2 \approx (k_t d/2)$

Anti-symmetric MSFVW for $kd \ll 1$: $-1 \approx (k_t d/2)$.

First, it is obvious that anti-symmetric MSFVW modes do not exist for $kd \ll 1$ since $k_t = \alpha$ should be positive. In contrast, the lowest-order symmetric mode is always excited, having an axial wavenumber k_{zi} with an order of magnitude larger than that of the transverse k_t one.

Concerning the frequency of this mode, we take the limit of (3.106) for $\alpha d \to 0$ (note that the second term requires the application of De L'Hospital's rule), which yields $\omega(n = 0 \to \omega_i$. That is, for very thin slabs (e.g., films) the lowest-order mode is a symmetric one, primarily z-directed or transverse to the film $\bar{k} \approx k_{zi} \cdot \hat{z}$ having $\omega \to \omega_i$. This is also confirmed in [17].

The other limiting case occurs when slab thickness becomes very large $kd >> 1$ and is likely to approach the plane-wave propagation in an infinite medium. This is first verified by taking the limit $\alpha d >> 1$ in (3.106), which yields to $\omega(n = 0) \to \sqrt{\omega_i(\omega_i + \omega_m)}$. This is identical to the plane wave case in (3.81) for $\theta = 0°$, which means that the $n = 0$ symmetric mode propagates in the z-direction even for very thick slabs. However, a very thick slab supports an infinite number of MSFVW, both symmetric and anti-symmetric.

3.22 LONGITUDINALLY MAGNETIZED INFINITELY EXTENDING THIN SLABS

Damon and Eshbach [13], originally studied the magnetostatic modes of a thin ferrite slab, magnetized to saturation along its plane. Once again, the coordinate system is modified in order to preserve the notation of the previous paragraphs so that the analysis can be covered by the Walker's theory. For this purpose, the x-axis is oriented perpendicularly to the slab plane as shown in Figure 3.15 and the DC biasing field, as well as saturation magnetization, is again aligned along the $\hat{z}$-axis.

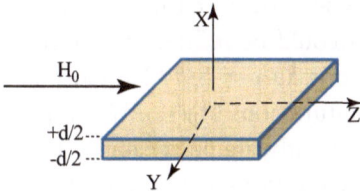

Figure 3.15: Infinitely extending thin ferrite slab longitudinally biased to saturation.

Firstly, in order to examine the internal DC magnetic field H_i recall that the demagnetization factors parallel to the slab are zero $N_y = N_z = 0$ while normal to that is unity, $N_x = 1$. However, only N_z is involved since $\hat{z}$-bias is assumed. Therefore, $\bar{H}_i = \bar{H}_0 - N_z \bar{M}_s = \bar{H}_0$ or $\omega_i = \mu_0 \gamma H_i = \omega_0$ unless the anisotropy field is accounted for in a way like that of Eq. (3.66).

The separation of variables according to Eqs. (3.82) to (3.85) can be employed again, but the orientation of the different coordinate needs to be taken into account. Following the same reasoning as in the previous section, the boundary conditions for the coordinates parallel to the slab surface require the same longitudinal eigenfunctions inside and outside the slab, that is,

$g_i(y) = g_e(y) = g(y)$ and $h_i(z) = h_e(z) = h(z)$. Therefore, Eq. (3.82) reads as follows:

$$\psi^i = f_i(x)g(y)h(z) \tag{3.113a}$$
$$\psi^e = f_e(x)g(y)h(z). \tag{3.113b}$$

Along the x-direction, normal to the slab, the solution (now denoted $f(x)$) must be like that of the normally biased case, that is, exponentially decaying outside the slab, which means that it vanishes at infinity ($x \to \pm\infty$) and has a standing wave form (co-sinusoidal or sinusoidal) inside the slab. Hence, in a way similar to that of Eqs. (3.89) and (3.91) we may write

$$f_i(x) = A_x \sin(k_{xi}x) + B_x \cos(k_{xi}x) \quad \text{for} \quad -d/2 \le x \le d/2 \tag{3.114}$$

$$f_e(x) = \left\{ \begin{array}{ll} \psi_{0e}e^{jk_{xe}(x-\frac{d}{2})} = \psi_{0e}e^{-\alpha(x-\frac{d}{2})} & \text{for} \quad x \ge d/2 \\ \psi'_{0e}e^{jk_{xe}(x+\frac{d}{2})} = \psi'_{0e}e^{+\alpha(x+\frac{d}{2})} & \text{for} \quad x \le -d/2 \end{array} \right\}. \tag{3.115}$$

Note that $k_{xe}^2 = -\alpha^2 < 0$ or $k_{xe} = j\alpha$.

The wavenumbers involved in the eigenfunctions $g(y)$ and $h(z)$ parallel to the slab, are identical inside and outside the slab. Also, must be $k_{ye} = k_{yi} = k_y$ and $k_{ze} = k_{zi} = k_z$ since the boundary conditions at $x = \pm d/2$ must hold at any arbitrary (y, z) point. The common practice is to consider solutions for g and h that represent waves propagating at arbitrary directions along $-\hat{y}$, $-\hat{z}$, like $e^{-j\bar{k}_t \cdot \bar{\rho}}$, where $\bar{k}_t = k_y\hat{y} + k_z\hat{z}$ and $\bar{\rho} = y\hat{y} + z\hat{z}$. This is because the slab is infinitely extending in the above directions. A specific direction in the y–z plane is defined by the ratio of k_y/k_z, which for an infinitely extending slab could be arbitrary, allowing propagation to occur in all directions along the plane. However, this statement is not general enough. Obviously, it holds for isotropic slabs, but it could be refuted in anisotropic materials. The question here is whether the boundary conditions at $x = \pm d/2$ impose any restrictions on the k_y and k_z values and the corresponding eigenfunctions $g(y)$ and $h(z)$ through the anisotropy of the ferrite's susceptibility tensor $[\mathcal{X}]$ or, equivalently, its permeability tensor $[\mu]$. To answer this question, let us first examine the boundary conditions at $x = \pm d/2$.

(i) Continuity of the tangential magnetic field h_y, h_z at $x = \pm d/2$:

$$h_y = -\frac{\partial\psi}{\partial y} \quad \leftrightarrow \quad f_i(x)\frac{dg(y)}{dy}h(z)\big|_{x=\pm d/2} = f_e(x)\frac{dg(y)}{dy}h(z)\big|_{x=\pm d/2} \tag{3.116a}$$

$$h_z = -\frac{\partial\psi}{\partial z} \quad \leftrightarrow \quad f_i(x)g(z)\frac{dh(z)}{dz}\bigg|_{x=\pm d/2} = f_e(x)g(z)\frac{dh(z)}{dz}\bigg|_{x=\pm d/2}. \tag{3.116b}$$

The intuitive selection of identical $g(y)$ and $h(z)$ eigenfunctions inside and outside the slab is required (or verified) by (3.116a) and (3.116b), which in view of this reduces to:

$$f_i(x)|_{x=\pm d/2} = f_e(x)|_{x=\pm d/2}. \tag{3.117}$$

(ii) Continuity of the normal flux density component $b_n = b_x$, which according to (3.72) and (3.73) is:

$$b_x = \begin{cases} b_{xi} \left((1 + \mathcal{X}) \cdot h_{xi} + j\mu_0 k_1 \cdot h_{yi} \right) & \text{in the slab} \quad |x| \le \frac{d}{2}) \\ b_{xe} = \mu_0 \cdot h_{xe} & \text{outside the slab} \quad |x| \ge \frac{d}{2}. \end{cases} \quad (3.118)$$

At this point, it is interesting to take a closer look at (3.72) and (3.73) to gain some intuitive understanding of the consequences of the different orientation of the $(\hat{x}, \hat{y}, \hat{z})$-axes with respect to the slab. First, when the $\hat{z}$-axis is normally oriented, the $\hat{x}$ and $\hat{y}$-axes retain a specific symmetry, both geometrically and in terms of the permeability anisotropy. Thus, in this case, a wave is expected to propagate in both $\hat{x}$ and $\hat{y}$ directions. In contrast, when $\hat{x}$ (or $\hat{y}$)-axis is oriented normal to the slab, the $\hat{y}$ (or $\hat{x}$) and $\hat{z}$-axes are geometrically symmetric, but there is an anisotropy along $\hat{y}$ (or $\hat{x}$), while the material is isotropic along the $\hat{z}$-axis. This can be justified in this case through Eq. (3.118), which involves $h_{yi} = -\partial \psi_i / \partial y$, but not $h_{zi} = -\partial \psi_i / \partial z$. Moreover, remember that this anisotropy caused the "self-guidance" effect observed in the case of magnetostatic modes propagating in an infinite (unbounded) medium; in other words, each mode propagates along a specific direction.

To return to the topic under discussion, Eq. (3.118) can be recast in terms of the potential functions as:

$$\left\{ (1 + \mathcal{X}) \frac{\partial \psi_i}{\partial x} + jk_1 \frac{\partial \psi^i}{\partial y} \right\}_{x=\pm d/2} = \frac{\partial \psi^e}{\partial x} \bigg|_{x=\pm d/2}. \quad (3.119a)$$

In view of (3.113), the above equation reads as follows:

$$\left\{ (1 + \mathcal{X}) \frac{df_i(x)}{dx} g(y) + jk_1 f_i(x) \frac{dg(y)}{dy} \right\}_{x=\pm d/2} = \frac{df_e(x)}{dx} g(y) \bigg|_{x=\pm d/2}. \quad (3.119b)$$

Note that the eigenfunction $h(z)$ is absent from (3.119b) since it is common in all terms. Thus, it could be arbitrarily selected. For example, the choice of Damon and Eshbach [13] was $h(z) = \cos(k_z z)$, but a propagating wave $e^{-jk_z z}$ could have been chosen as well. However, a closer examination of conditions (3.115) at $x = +d/2$ and $x = -d/2$ reveal that $g(y)$ can only be chosen as a propagating wave, e.g., $g(y) = e^{-jk_y \cdot y}$. This is also stated in [13], and a similar condition is observed by Walker [9], who found that the only possible azimuthal dependence in the spheroid case is $e^{jm\varphi}$. Moreover, in order to be compatible with [13], a positive sign in the exponential is assumed (e.g., propagation in the negative y-direction). Thus,

$$g(y) = e^{jk_y y} \quad (3.120)$$

which yields

$$\frac{dg}{dy} = +jk_y \cdot e^{jk_y y} = jk_y \cdot g(y) \quad \text{or} \quad \frac{d}{dy} \rightarrow jk_y. \quad (3.121)$$

In view of the latter, $g(y)$ may also be eliminated from (3.119b), which now reads as follows:

$$\left\{(1 + \mathcal{X})\frac{df_i(x)}{dx} - k_1 \cdot k_y \cdot f_i(x)\right\}_{x=\pm d/2} = \frac{df_e(x)}{dx}. \tag{3.122}$$

In turn, Eqs. (3.122) and (3.117) must be combined with the wavenumber conditions (3.85) resulting for the separation of variables. For the present case, the latter can be rewritten as:

$$(1 + \mathcal{X})k_{xi}^2 + (1 + \mathcal{X})k_y^2 + k_z^2 = 0 \tag{3.123a}$$
$$-\alpha^2 + k_y^2 + k_z^2 = 0. \tag{3.123b}$$

Substituting (3.123b) into (3.123a) yields

$$(1 + \mathcal{X})k_{xi}^2 + \mathcal{X} \cdot k_y^2 + \alpha^2 = 0. \tag{3.123c}$$

Let us now enforce the boundary conditions. Starting from the tangential field components, (3.117) through (3.114) and (3.115) read as follows:

$$x = +d/2: \quad A_x \sin(k_{xi}d/2) + B_x \cos(k_{xi}d/2) = \Psi_{0e} \tag{3.124a}$$
$$x = -d/2: \quad -A_x \sin(k_{xi}d/2) + B_x \cos(k_{xi}d/2) = \psi'_{0e}. \tag{3.124b}$$

These can be solved for A_x and B_x as:

$$A_x = \frac{\psi_{0e} - \psi'_{0e}}{2} \cdot \frac{1}{\sin(k_{xi}d/2)} \tag{3.125a}$$
$$B_x = \frac{\psi_{0e} + \psi'_{0e}}{2} \cdot \frac{1}{\cos(k_{xi}d/2)}. \tag{3.125b}$$

The procedure can be simplified by substituting A_x and B_x back into (3.114) to yield

$$f_i(x) = \frac{\psi_{0e} - \psi'_{0e}}{2} \cdot \frac{\sin(k_{xi}x)}{\sin(k_{xi}d/2)} + \frac{\psi_{0e} + \psi'_{0e}}{2} \cdot \frac{\cos(k_{xi}x)}{\cos(k_{xi}d/2)}. \tag{3.126}$$

It is interesting to observe $f_i(x)$ at $x = \pm d/2$, which is

$$f_i(x = +d/2) = \psi_{0e} \quad \text{and} \quad f_i(x = -d/2) = \psi'_{0e}. \tag{3.127}$$

The boundary condition (3.122) for the normal component can be written explicitly through (3.115) and (3.126) as:

$x = +d/2$:

$$(1 + \mathcal{X})k_{xi}\left\{\frac{\Psi_{0e} - \Psi'_{0e}}{2}\cot\left(k_{xi}\frac{d}{2}\right) - \frac{\Psi_{0e} + \Psi'_{0e}}{2}\tan\left(k_{xi}\frac{d}{2}\right)\right\} - k_1 k_y \cdot \Psi_{0e} = -\alpha \cdot \Psi_{0e}$$

(3.128a)

$x = -d/2$:

$$(1 + \mathcal{X})k_{xi}\left\{\frac{\Psi_{0e} - \Psi'_{0e}}{2}\cot\left(k_{xi}\frac{d}{2}\right) + \frac{\Psi_{0e} + \Psi'_{0e}}{2}\tan\left(k_{xi}\frac{d}{2}\right)\right\} - k_1 k_y \cdot \Psi'_{0e} = \alpha \cdot \Psi'_{0e}.$$

(3.128b)

Some algebraic manipulation is required to eliminate Ψ_{0e} and Ψ'_{0e} from (3.128a) and (3.128b). For example, in each equation, all terms containing Ψ_{0e} can be moved to the left side, and all terms with Ψ'_{0e} to the right side. Then, by dividing the two equations, we obtain the modes characteristic equation:

$$-(1 + \mathcal{X})^2 k_{xi}^2 + (1 + \mathcal{X}) \cdot k_{xi}\alpha\left[\cot\left(k_{xi}\frac{d}{2}\right) - \tan\left(k_{xi}\frac{d}{2}\right)\right] + \alpha^2 - k_1^2 k_y^2 = 0. \quad (3.129)$$

For further simplification, recall the trigonometric identity:

$$\tan(2z) = 2/[\cot(z) - \tan(z)] = 1/\cot(2z). \quad (3.130)$$

Then (3.129) becomes

$$2(1 + \mathcal{X}) \cdot k_{xi} \cdot \alpha \cdot \cot(k_{xi}d) - (1 + \mathcal{X})^2 \cdot k_{xi}^2 + \alpha^2 - k_i^2 k_y^2 = 0. \quad (3.131)$$

Equation (3.131) is identical to Damon and Eshbach's [13], even though different symbols are used herein. The characteristic equation (3.131) along with conditions (3.123) determines the mode spectrum. As in the previous section, the quantity $(1 + \mathcal{X})$ may be either positive or negative. Recall (3.100) as:

$$1 + \mathcal{X} < 0 \quad \text{for} \quad \omega_i < \omega < [\omega_i(\omega_i + \omega_m)]^{1/2}. \quad (3.132a)$$

$$1 + \mathcal{X} > 0 \quad \text{for} \quad \omega < \omega_i \quad \text{or} \quad \omega > [\omega_i(\omega_i + \omega_m)]^{1/2}. \quad (3.132b)$$

In addition, from the definition of $\mathcal{X}$ in (3.90) it is obvious that

$$\mathcal{X} > 0 \text{ for } \omega < \omega_i, \quad \mathcal{X} \rightarrow \pm\infty \text{ at } \omega = \omega_i, \text{ and } \mathcal{X} < 0 \text{ for } \omega > \omega_i.$$

Note that the infinity of $\mathcal{X}$ at $\omega = \omega_i$ is a direct consequence of ignoring ferrite losses, while the inclusion of losses results in a finite value. The behavior of both $(1 + \mathcal{X})$ and $\mathcal{X}$ vs. frequency is presented in Figure 3.16.

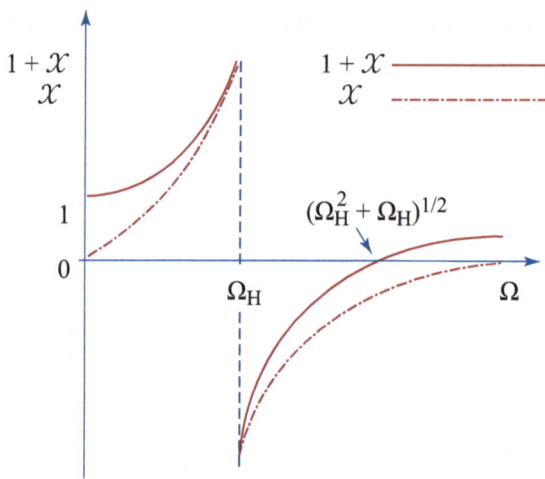

Figure 3.16: $\mathcal{X}$ and $(1 + \mathcal{X})$ vs. circular frequency [13].

For convenience, let us adopt the notation of [13] for the normalized frequencies as $\Omega = \omega/\omega_m$ and $\Omega_H = \omega_i/\omega_m$. In turn, observing Figure 3.16, three frequency ranges may be identified for the signs of $\mathcal{X}$ and $(1 + \mathcal{X})$:

$$\text{Range I:} \qquad \Omega < \Omega_H \leftrightarrow \mathcal{X} > 0, \ (1 + \mathcal{X}) > 0 \tag{3.133a}$$

$$\text{Range II:} \qquad \Omega_H < \Omega < \sqrt{\Omega_H^2 + \Omega_H} \leftrightarrow \mathcal{X} < 0, (1 + \mathcal{X}) < 0 \tag{3.133b}$$

$$\text{Range III:} \qquad \Omega > \sqrt{\Omega_H^2 + \Omega_H} \leftrightarrow \mathcal{X} < 0, (1 + \mathcal{X}) > 0. \tag{3.133c}$$

Moreover, recall that the slab is infinitely extended in both the y and z directions, we expect K_y and K_z to be real (either positive or negative) and to have continuous characteristic values [13]. Furthermore, if we restrict ourselves to surface-wave behavior in the air region above or below the slab, then the field should be attenuated exponentially in the x-direction tending to zero for $x \to +\infty$ or $x \to -\infty$. In this case, $\alpha = k_{xe} > 0$; that is, α should be real and positive. In view of the above and ignoring ferrite losses for clarification convenience, K_{xi} could be either real or imaginary, or, correspondingly, $K_{xi}^2 > 0$, $K_{xi}^2 < 0$. The former ($K_{xi}^2 > 0$) could be identified as a magnetostatic volume wave since energy spreads all over the sample (and as a standing wave in the x-direction). The examination of phase and group velocities will be shown next to have opposite signs and for this reason this wave should be termed "backward" and be symbolized as MSBVW (magneto-static backward volume wave).

In contrast, when $K_{xi}^2 < 0$ or K_{xi} is imaginary, the field is exponentially damped in the x-direction (normal to the slab or film) inside the ferrite slab as well as outside it (it has already been noted that α is real and positive, yielding an $e^{-\alpha|x-d|}$ dependence). That is, energy is concentrated along the top or the bottom of the ferrite-air interface and propagates along it (in the y- and z-directions). Therefore, these types of modes are termed "surface modes" and symbolized as MSSW (magneto-static surface waves).

The question now is which type of mode is excited in each of the three regions identified in (3.133) and what their specific characteristics are. For range I as well as range III defined in (3.133), it is obvious that the only requirement that needs to be met for (3.123a) or (3.123c) is for $k_{xi}^2 < 0$ or k_{xi} to be imaginary, so these modes should be MSSW. In contrast, in range II, the negative sign of $(1 + \mathcal{X})$ enables the choice $k_{xi}^2 > 0$, so MSBVW are supported. In addition, remember that range II is the extrapolated spin-wave band since it was previously found that this is the only range supporting spin waves provided that the already mentioned wavelength restrictions are met. Therefore, outside this range k_{xi} must be imaginary [13].

3.22.1 MAGNETOSTATIC VOLUME MODES $(1 + \mathcal{X}) < 0$

As already noted, in range II, it is $k_{xi}^2 > 0$. This is a consequence of $(1 + \mathcal{X}) < 0$. To examine the characteristics, we may express the wavenumber in spherical coordinates as in [13]:

$$\text{Ferrite region:} \quad \bar{k}_i = k_{xi}\hat{x} + k_y\hat{y} + k_z\hat{z} \tag{3.134a}$$

$$\text{Air region:} \quad \bar{k}_\alpha = k_{xe}\hat{x} + k_y\hat{y} + k_z\hat{z} \quad \text{where} \quad k_{xe} = -j\alpha \tag{3.134b}$$

$$\text{and} \quad k_{xi} = k_{ti}\cos\varphi, \quad k_y = k_{ti}\sin\varphi, \quad k_{ti} = k_i\sin\theta, \quad k_z = k_i\cos\theta, \tag{3.134c}$$

where θ and φ are the angles of $\bar{k}_i$ with respect to the $\hat{z}$ and $\hat{x}$ axes, respectively.

In view of the above notation, (3.123b) reads as follows:

$$k_\alpha^2 = -\alpha^2 + k_y^2 + k_z^2 = 0. \tag{3.135a}$$

This simply means that the total wavenumber in the air region is zero.

In turn, (3.123a) may be written in two similar forms:

$$(1 + \mathcal{X})k_{ti}^2 + k_z^2 = 0 \quad \Leftrightarrow \quad \frac{k_{ti}^2}{k_z^2} = \tan^2\theta = \frac{-1}{(1 + \mathcal{X})} \tag{3.135b}$$

$$\text{or} \quad \mathcal{X} \cdot k_{ti}^2 + k_i^2 = 0 \quad \Leftrightarrow \quad \frac{k_{ti}^2}{k^2} = \sin^2\theta = -\frac{1}{\mathcal{X}}. \tag{3.135c}$$

Using the definition of $\mathcal{X}$ from (3.84), the latter yields

$$\sin^2\theta = \frac{\Omega^2 - \Omega_H^2}{\Omega_H} \quad \Leftrightarrow \quad \Omega^2 = \Omega_H^2 + \Omega_H\sin^2\theta. \tag{3.136}$$

Equation (3.136) reveals once again the frequency range of volume modes defined by $0 \leq \sin^2\theta \leq 1$. Their lower frequency corresponds to $\theta = 0°$, where $\Omega = \Omega_H$, $\bar{k}_i = k_z \cdot \hat{z}$ and

$k_{ti} = 0$, which means that the wave propagates along the direction of DC magnetization (z-axis). In contrast, $\theta = 90°$ corresponds to their maximum frequency $\Omega = [\Omega_H^2 + \Omega_H]^{1/2}$ and $k_z = 0, k_{ti} = k$. Since the mode pattern in the x-direction is that of a standing wave, the wave propagates in either positive or negative y-directions. Recall that both signs of k_y are permitted and a dependence of the form $e^{jk_y y}$ is assumed. Since these two modes are degenerate in frequency, they may be combined to form two new modes [13]. However, these combined modes propagate in opposite directions on the two sides of the slab. It is also interesting to observe the special case of modes appearing at $\theta = 0°$. From (3.123a), it can be seen that $\alpha = 0$. Thus, these modes do not have an RF field outside the slab. On the other hand, from (3.123a) we can see that k_{xi} is arbitrary since $(1 + \mathcal{X}) = 0$ at this frequency; in other words, these types of mode have no variation along the longitudinal dimensions (y, z) while their field pattern is arbitrary in the transverse x-direction. Moreover, it will be proved next that for $\theta = 90°$ these modes become modes of the surface wave type (MSSW).

Density of modes: Equations (3.123) clearly show that the number of possible volume modes is proportional to $\mathcal{X}$ and $(1 + \mathcal{X})$. These quantities tend to infinity, and so does the number of modes when $\Omega \to \Omega_H$. However, this phenomenon appears only at the ideal ferrite slab case when ferrite losses do not exist. In actual practice, when ferrite losses are included, $\mathcal{X}$ becomes finite at $\Omega \to \Omega_H$ and the number of possible modes is drastically reduced.

Phase and group velocities: The phase (V_p) and group velocities (V_g) are expressed in a way similar to that of Eq. (3.111), but here propagation occurs in both the $\hat{y}$ and $\hat{z}$ directions as $e^{jk_y y}e^{-jk_z z} = e^{-j\bar{k}_\rho \cdot \bar{\rho}}$, where $\bar{k}_\rho = -k_y \hat{y} + k_z \hat{z}$ and $\bar{\rho} = y\hat{y} + z\hat{z}$. Therefore, k_t in (3.111) should be replaced by k_ρ. This, in turn, is $k_\rho = \alpha$ from Eq. (3.123b). Thus, $V_p = \omega/\alpha$ and $\alpha => 0$ could be obtained from the solution of the dispersion equation (3.131). However, group velocity $V_g = (d\omega/dk_\rho)$ requires the differentiation of (3.131), which results in very complicated expressions. In contrast, relatively simple expressions clarifying the properties of the modes can be obtained for the particular cases when:

(i) $\theta = 0°$ and $k_y = 0$ and the wave propagates along the z-direction and

(ii) $\theta = 90°$ and $k_z = 0$ and the wave propagates in the y-direction.

Propagation in the z-direction: ($k_y = 0$ $\leftrightarrow$ MSBVW)
The phase and group velocities for this particular case read as follows:

$$V_p = \frac{\omega}{k_z} \quad \text{and} \quad V_g = \frac{\partial \omega}{\partial k_z}. \tag{3.137}$$

Equations (3.123) for $k_y = 0$ are reduced to:

$$\alpha = k_{xe} = k_z \quad \text{and} \quad k_{xi} = k_z/\sqrt{-(1 + \mathcal{X})}. \tag{3.138}$$

Substituting back into (3.131), the dispersion equation is simplified to:

$$2(1 + \mathcal{X})k_z^2 \cdot \sqrt{-(1 + \mathcal{X})} \cdot \cot\left(k_z d / \sqrt{-(1 + \mathcal{X})}\right) + (2 + \mathcal{X}) \cdot k_z^2 = 0. \qquad (3.139a)$$

Thus, one possible solution is $k_z = 0$, which has already been discussed as the case ($k_z = 0, k_y = 0$). By carefully handling $(1 + \mathcal{X}) < 0$ terms, the interesting solution reads as follows:

$$\cot\left(k_z d / \sqrt{-(1 + \mathcal{X})}\right) = +\frac{1}{2}(2 + \mathcal{X}) / \sqrt{-(1 + \mathcal{X})} = \frac{1}{\sqrt{-(1 + \mathcal{X})}} - \sqrt{-(1 + \mathcal{X})}.$$

$$(3.139b)$$

Note that $(1 + \mathcal{X})$ in the range of volume waves is negative. Also, the question of the cotangent and its argument appears. First, a function of the form $\left(\frac{1}{y} - y\right)$ with $y = -(1 + \mathcal{X}) > 0$ is obviously positive when $y < 1$, zero at $y = 1$ and negative for $y > 1$. Therefore, this is also the behavior of the cotangent in (3.139b). In turn, in range II volume waves are split into two sub-ranges at $-(1 + \mathcal{X}) = 1$. Using the definition (3.86) for $\mathcal{X}$, we have that:

$$-(1 + \mathcal{X}) = 1 \quad \leftrightarrow \quad \Omega = \Omega_A = \left[\Omega_H^2 + \frac{1}{2}\Omega_H\right]^{1/2}. \qquad (3.140a)$$

These sub-ranges can be set as follows:

Range II. a:

$$\Omega_H < \Omega = \sqrt{\Omega_H^2 + \frac{1}{2}\Omega_H} < \Omega_A \quad \leftrightarrow \quad (1 + \mathcal{X}) < -1 \text{ negative cotangent} \qquad (3.140b)$$

Range II. 0:

$$\Omega_A = \Omega \quad \leftrightarrow \quad (1 + \mathcal{X}) = -1 \text{ zero cotangent} \qquad (3.140c)$$

Range II. b:

$$\Omega_A < \Omega = \sqrt{\Omega_H^2 + \Omega_H} < \Omega_B \quad \leftrightarrow \quad -1 < (1 + \mathcal{X}) < 0 \text{ positive cotangent.} \qquad (3.140d)$$

The characteristic equation (3.139b) is very simple at $\Omega = \Omega_A$ and can be readily solved as follows:

$$\Omega = \Omega_A \quad \leftrightarrow \quad \cot\left(k_z d / \sqrt{-(1 + \mathcal{X})}\right) = 0$$

$$\leftrightarrow \quad k_z d / \sqrt{-(1 + \mathcal{X})} = n \cdot \frac{\pi}{2} \quad \text{where} \quad n = 1, 3, 5. \qquad (3.141)$$

Likewise, the argument of the cotangent falls in the second or fourth quadrant for range II.a and in the first and third ones for range II.b.

In turn, group velocity can be obtained by differentiating (3.139) with respect to k_z, but remember that $\mathcal{X}$ is a function of (ω) given in Eq. (3.86). However, instead of dealing with (3.139), it is preferable to transform it into a more familiar form. For this purpose, we may adopt the half argument identity like that of (3.130) and solve (3.139b) by assuming that $\tan(z) = 1/\cot(z)$ is the unknown. The resulting second-order polynomial equation has two distinct solutions:

$$\tan\left(\frac{k_z d}{2A}\right) = \pm\frac{1}{A}.$$

(3.142a)

The above can also be written as:

$$\cot\left(\frac{k_z d}{2A}\right) = \pm A,$$

(3.142b)

where $A = \sqrt{-(1 + X)}$.

Now, in view of the identity (3.107a), it can be clearly seen that (3.142b) represents the odd ($n = 1, 3, 5, \ldots$) and the even modes ($n = 0, 2, 4, \ldots$) of a dispersion equation of the form

$$\tan\left(\frac{k_z d}{2\sqrt{-(1 + \mathcal{X})}} - \frac{n\pi}{2}\right) = \sqrt{-(1 + \mathcal{X})} \quad n = 0, 1, 2, 3, \ldots$$

(3.143)

This dispersion equation is identical to the one given in [12], with a minor difference in the definition of (n). This can be solved graphically if it is recast in a way similar to that of (3.109) and (3.110). An approximate solution of (3.143) for the lowest-order mode ($n = 0$) is again given by Kallinikos [38]:

$$\omega^2 = \omega_i \left\{ \omega_i + \omega_m \left(\frac{1 - e^{-k_z d}}{k_z d}\right) \right\}.$$

(3.144)

The dispersion curves of $\Omega = \omega/\omega_m$ vs. $k_z d$ are shown in Figure 3.17 [12]. It is obvious that the slope of these curves is negative and group velocity V_g is also negative; hence, they are characterized as backward waves (MSBVW).

The mathematical expression for group velocity can be obtained from the differentiation of (3.144) with respect to k_z [12], as follows:

$$V_g = \frac{d\omega}{dk_z} = \frac{2k_1}{\omega_m d} + \frac{k_1 k_z}{\omega_m} \cdot \frac{\mathcal{X}}{1 + \mathcal{X}}.$$

(3.145a)

The lowest-order ($n = 0$) mode has the steepest slope and, thus, the most negative group velocity. For this mode and for electrically thin slabs or films ($k_z d \ll 1$), the above expression can be approximated to

$$\left.\frac{1}{V_g}\right|_{k_z d \ll 1} \approx -\frac{4}{\omega_m d} \cdot \frac{\sqrt{\omega_i (\omega_i + \omega_m)}}{\omega_i}.$$

(3.145b)

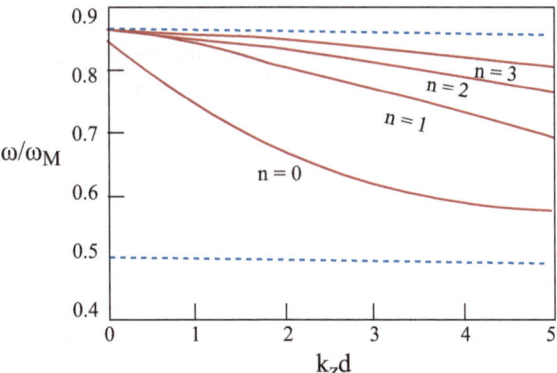

Figure 3.17: Dispersion curves for backward magnetostatic volume waves (MSBVW) [12].

It is obvious from (3.145b) that group velocity is negative; hence, the modes are correctly termed "backward waves." The inverse of group velocity $(1/V_g)$ vs. $\Omega = \omega/\omega_m$ for the first four modes is presented in Figure 3.18 [12].

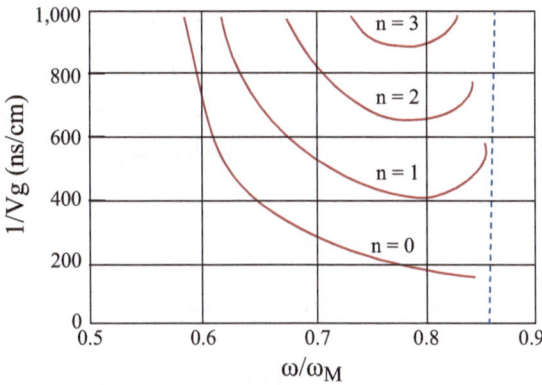

Figure 3.18: Inverse group velocity $(1/Vg)$ for MSBVW [12].

Before moving on to the analysis of magnetostatic surface waves it is worth examining whether volume modes are possible when pure propagation only in the y-direction is assumed. That is, consider $k_z = 0$ and $\bar{k} = k_{xi}\hat{x} + k_y\hat{y}$. In this case, Eq. (3.123a) reads as follows:

$$(1 + \mathcal{X})k_{xi}^2 + (1 + \mathcal{X})k_y^2 = 0 \quad \leftrightarrow \quad k_{xi}^2 = -k_y^2 \quad \leftrightarrow \quad k_{xi} = \pm jk_y. \quad (3.145c)$$

It is then obvious that k_{xi}^2 should be negative or k_{xi} should be imaginary when $k_z = 0$ since pure propagation in the y-direction requires k_y to be real. However, this is the case of magnetostatic *surface* waves, (to be discussed next), as the imaginary k_{xi} causes the field to decay very fast (almost exponentially) toward the interior of the slab away from either surface $x = +d/2$ or $-d/2$. Consequently, volume waves cannot propagate purely transversely to the direction of DC magnetization (z-axis). Only surface modes can do that.

It is again interesting to examine the MSBVW characteristics with respect to slab thickness. The above proves that volume waves are not supported for pure transverse propagation. For the infinite thickness limit $kd >> 1$, Eq. (3.144), obtained for $k_y = 0$ and describing the lowest order mode ($n = 0$), gives $\omega \to \omega_i$ or $\Omega \to \Omega_H$. Again, this approaches the plane-wave limit and the wave is purely propagating in the z-direction (along DC magnetization). In contrast, in the electrically thin slab (or film), where $kd << 1$, propagation is almost transverse $k_z \to 0$, and this occurs at the maximum frequency $\Omega \to [\Omega_H^2 + \Omega_H]^{1/2}$ or $\omega \to [\omega_i^2 + \omega_i \omega_m]^{1/2}$.

3.23   MAGNETOSTATIC SURFACE WAVES $(1 + \mathcal{X}) > 0$

The above analysis as well as the original work of [13] shows that outside range II, where $(1 + \mathcal{X}) < 0$, the only possible modes are those of the surface-wave type. This is supported by the requirement that $k_{xi}^2 < 0$ or that k_{xi} be imaginary to obey the dispersion relations. In view of this, one may set k_{xi} equal to

$$k_{xi} = j\alpha_i \quad \text{or} \quad k_{xi}^2 = -\alpha_i^2. \tag{3.146}$$

The analysis for volume modes is still applicable as long as the argument of the functions is allowed to be complex. A more appropriate characteristic equation for surface waves in ranges I and II is obtained by substituting (3.146) into the previous expressions. First, the x-dependence of Eq. (3.114) may be rewritten as:

$$f_i(x) = A'_x e^{+\alpha_i x} + B'_x e^{-\alpha_i x}. \tag{3.147}$$

Writing (sinh) into an exponential form we have

$$A'_x = \frac{B_x + jA_x}{2} = \frac{1}{2\sinh(\alpha_i d)} \cdot \left(\psi_{0e} e^{\alpha_i d/2} - \psi'_{0e} e^{-\alpha_i d/2}\right) \tag{3.148a}$$

$$B'_x = \frac{B_x - jA_x}{2} = \frac{1}{2\sinh(\alpha_i d)} \cdot \left(\psi'_{0e} e^{\alpha_i d/2} - \psi_{0e} e^{-\alpha_i d/2}\right). \tag{3.148b}$$

Substituting the above coefficients back into (3.147) yields

$$f_i(x) = \frac{1}{\sinh(\alpha_i d)} \cdot \left(\psi_{0e} \cdot \sinh[\alpha_i(x + d/2)] - \psi'_{0e} \cdot \sinh[\alpha_i(x - d/2)]\right). \tag{3.149}$$

Alternatively, the same expression can be obtained by substituting (3.146) directly into (3.126). Moreover, the characteristic equation (3.129) reads as follows:

$$2(1 + \mathcal{X}) \cdot \alpha \cdot \alpha_i \cdot \coth(\alpha_i d) + (1 + \mathcal{X})^2 \cdot \alpha_i^2 + \alpha^2 - k_i^2 k_y^2 = 0. \tag{3.150}$$

This is similar to that given in [38] even though a different notation and axis orientation is used therein.

In order to clarify the characteristics of these modes, let us examine the particular case in which $k_z = 0$, that is, when the wave propagates only in the y-direction (transverse to DC-magnetization).

3.23.1   MSSW PROPAGATING IN THE y–DIRECTION ($k_z = 0$)

In view of the simplification $k_z = 0$, expressions (3.123) reduce to

$$\alpha = |k_y| \quad \text{or} \quad k_y = \pm\alpha \quad \text{and} \quad \alpha_i^2 = k_y^2 \quad \text{or} \quad k_y = \pm\alpha_i. \tag{3.151}$$

Recall that k_y may be either positive or negative, while α is real and positive in order to represent exponential attenuation in the air region. In turn, the characteristic equation (3.150) is simplified to:

$$2(1 + \mathcal{X}) \coth(\alpha_i d) + (1 + \mathcal{X})^2 + 1 - k_i^2 = 0. \tag{3.152}$$

The trivial mathematical solution $\alpha_i = 0$ yields $k_y = 0$, $\alpha = 0$ and $k_z = 0$, but also inspecting (3.149) it represents zero field. Considering expressions (3.71) for $\mathcal{X}$ and k_i, the characteristic equation (3.150) may be conveniently rewritten as:

$$\Omega^2 = \Omega_H^2 + \Omega_H + [2 + 2\coth(\alpha_i d)]^{-1}. \tag{3.153}$$

Again, this is like that of [13], but for $\alpha_i \rightarrow |k_y|$. To examine (3.153) just recall the properties of function $\coth(z)$ (e.g., Abramowitz and Stegun [39]), where $\coth(z) \geq 1$ or $\coth(z) \leq -1$.

The first case yields the following upper frequency limit for the excitation of magnetostatic surface waves (MSSW):

$$\coth(\alpha_i d) \geq 1 \quad \leftrightarrow \quad \Omega \geq \left[\Omega_H^2 + \Omega_H + \frac{1}{4}\right]^{1/2} = \left(\Omega_H + \frac{1}{2}\right). \tag{3.154}$$

The other option, $\coth(\alpha_i d) \leq 1$, is not acceptable since it requires $\alpha_i < 0$, and this should also be valid for k_y as $\coth(k_y d) \leq 1$. Moreover, the $\coth(z)$ function is non-periodic; thus, there will be only one possible solution for each specific $\mathcal{X}$, namely, for each specific DC biasing of the ferrite film. In addition, this solution will only be valid in range III, defined in (3.133c) for frequencies above the range of volume modes. It is also clear that surface modes cannot be excited in range I. The only possible mode can be explicitly obtained through an analytical solution of (3.153), which, according to [12], reads as follows:

$$\omega^2 = \omega_i \left(\omega_i + \omega_m\right) + \frac{\omega_m^2}{4}\left[1 - e^{-2|k_y|d}\right]. \tag{3.155}$$

Expression (3.155) can be defined from (3.152) by considering the following identity:

$$\coth^{-1}(A) = \frac{1}{2}\ln\left\{(A + 1)/(A - 1)\right\} \quad \text{for} \quad A^2 > 1. \tag{3.156}$$

For $A = \mathcal{X}$.

In turn, the solution of (3.146) for coth gives

$$\coth{(\alpha_i d)} = \coth{(|k_y|d)} = \left\{ \frac{1}{2\left(\Omega^2 - \Omega_\beta^2\right)} - 1 \right\}, \tag{3.157a}$$

where $\Omega_\beta^2 = \Omega_H^2 + \Omega_H$.

Let us suppose that $\mathcal{X}$ is equal to the right side of (3.157a) and exploit (3.156) to obtain

$$|k_y|d = -\frac{1}{2}\ln{\left(1 - 4\Omega^2 + 4\Omega_\beta^2\right)}. \tag{3.157b}$$

Taking the antilogarithm and using $\Omega = \omega/\omega_m$ and $\Omega_H = \omega_i/\omega_m$ we go to expression (3.155). Group velocity can again be obtained through the differentiation of (3.155) with respect to k_y. Thus,

$$V_g = \frac{d\omega}{dk_y} = \frac{\omega_m^2 d}{4\omega}e^{-2|k_y|d}. \tag{3.158}$$

For electrically thin slabs or films, $|k_y|d \ll 1$, the above expression can be simplified to

$$\left.\frac{1}{V_g}\right|_{|k_y|d \ll 1} \approx \frac{4}{\omega_m^2 d}\sqrt{\omega_i\left(\omega_i + \omega_m\right)}. \tag{3.159}$$

The dispersion relation (3.155) is plotted in Figure 3.19. In this figure, ω/ω_m vs. $|k_y|d$ is given. Moreover, the inverse of group velocity ($1/V_g$) vs. ω/ω_m is presented in Figure 3.20.

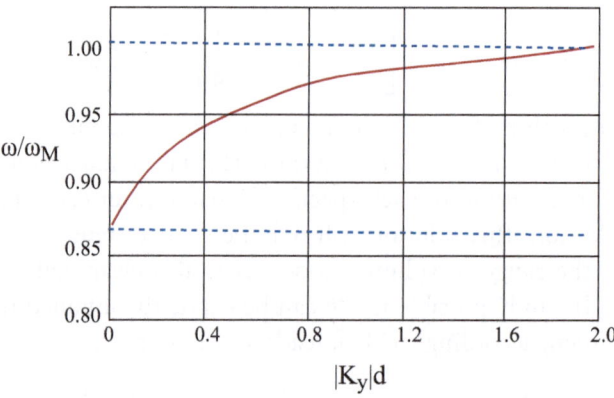

Figure 3.19: Dispersion relation for MSSW in a longitudinally magnetized film [12].

The magnetic potential profile, $f_i(x)$, in a direction transverse to the film is indicatively shown in Figure 3.21.

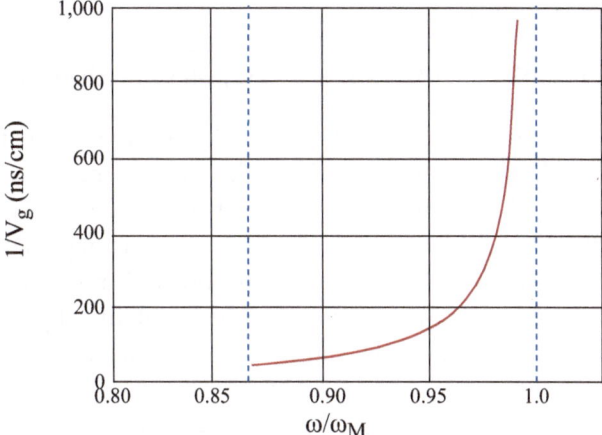

Figure 3.20: Inverse group velocity of MSSW [12].

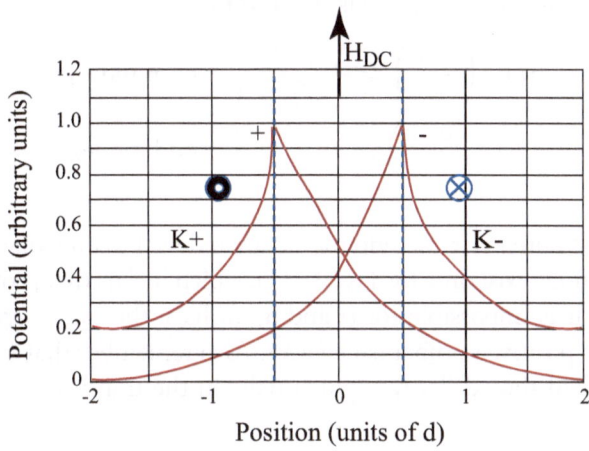

Figure 3.21: The magnetic potential $f_i(x)$ distribution for MSSW modes in a direction transverse to the film.

Figure 3.21 reveals that any increase in $\alpha_i = |k_y|$ causes the mode amplitude to diminish toward the film's interior at a higher rate; in other words, as $|k_y|$ increases, the MSSW modes cling more closely to the surface of the slab [13]. In addition, note that (3.123a) gives

$$\alpha_i^2 = k_y^2 + k_z^2/(1 + \mathcal{X}). \tag{3.160}$$

Therefore, when $k_z \neq 0$ and for the same k_y, it is obvious that α_i becomes larger and the MSSW mode clings more tightly to the surface. Furthermore, in the limit when $(1 + \mathcal{X}) \rightarrow 0^+$, the decay rate α_i tends to infinity:

$$(1 + \mathcal{X}) \to 0^+ \quad \text{or} \quad \Omega \to \Omega_\beta = \sqrt{\Omega_H^2 + \Omega_H} \quad \text{then} \quad \alpha_i \to 0. \tag{3.161}$$

Let us now elaborate on mode characteristics vs. slab thickness. For an electrically thin slab or film, $|k_y|d << 1$, by taking the limit in (3.155), the MSSW modes occur at $\omega \to [\omega_i (\omega_i + \omega_m)]^{1/2}$ or $\Omega \to [\Omega_H^2 + \Omega_H]^{1/2}$; that is, they occur at their minimum possible frequency. In contrast, for electrically thick slabs, these modes occur at their maximum possible frequency $\omega \to (\omega_i + \frac{1}{2}\omega_m)$ or $\Omega \to (\Omega_H + \frac{1}{2})$, as can be easily observed from (3.155).

To conclude this section, it is worth examining the possibility of purely longitudinal propagating surface waves. For this purpose, by substituting $k_y = 0$ and $k_{xi} = j\alpha_i$ into (3.121), we get:

$$k_z^2 = -(1 + \mathcal{X})k_{xi}^2 = (1 + \mathcal{X})\alpha_i^2 \to k_z = \pm\alpha_i \sqrt{1 + \mathcal{X}} \quad \text{and} \quad k_z^2 = \alpha^2. \tag{3.162}$$

The characteristic equation for $k_y = 0$ has already been given in (3.142) for volume-wave modes. With the above substitutions, this is reduced to:

$$\tanh\left(\frac{\alpha_i d}{2}\right) = \mp \frac{1}{\sqrt{1 + \mathcal{X}}} \quad \text{even modes} \tag{3.163a}$$

$$\coth\left(\frac{\alpha_i d}{2}\right) = \mp\sqrt{1 + \mathcal{X}} \quad \text{odd modes} \tag{3.163b}$$

Observe first that both even and odd modes are reduced to only one surface mode. Recall that surface-wave solutions exist only for $1 + \mathcal{X} > 0$, which means that $|\coth(z)| \geq 1$. Although hidden in the details, it is impossible for α and α_i to have the same sign because this would require that (coth) of a positive number is negative. Also remember that $k_{xe} = j\alpha$ and α should be positive since the field should decay exponentially in the transverse x-direction away from the slab.

3.23.2   NON–RECIPROCAL SURFACE–WAVE MODES

The analysis in the previous section shows that both signs of k_y are possible for MSSW. Thus, surface-wave modes may be excited in pairs at the same frequency or, as it is commonly said, these pairs of modes are degenerative in frequency. A particularly interesting phenomenon has been discovered since the original work of Damon and Eshbach concerning the non-reciprocal behavior of MSSW, explained as follows. The standing waves in the transverse x-direction lead to unequal amplitudes $\Psi_{0e} \neq \Psi'_{0e}$ at the top and bottom surfaces of the slab. Reversal of the sign of k_y reverses the direction of propagation and reflects the transverse field pattern $f_i(k)$, given in Eq. (3.149) in $x = 0$ plane [13]. The result is the same if the slab is rotated 180° around the x-axis. Moreover, since these two modes are degenerative in frequency, they appear simultaneously. One of them clings onto the top surface ($x = +d/2$), propagating toward the negative

y-direction as $e^{+jk_y y}$ [12]. In contrast, the mode with the opposite sign clings onto the bottom surface $(x = -d/2)$ and propagates toward the positive y-direction as $e^{-jk_y y}$. From a different point of view, this pair behaves like a single mode, with wavenumber k_y in the y-direction, which propagates to the end as the film on one surface and comes back propagating on the other surface. Note that a quite similar phenomenon occurs at high microwave frequencies, where an electromagnetic mode clings to either edge of a ferrite slab, propagating in opposite directions. Because of this property, these modes are called "edge modes." The mode configuration for short-wavelength surface waves with $k_z = 0$ is given in Figure 3.22.

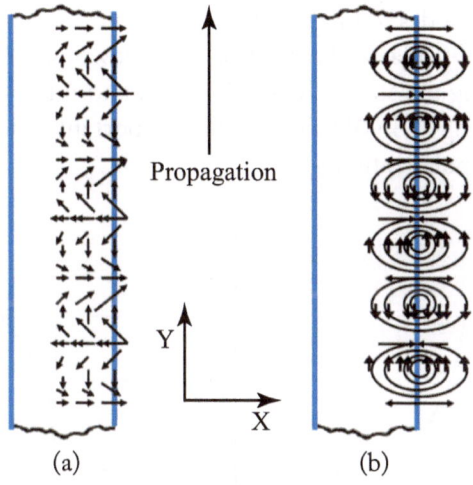

(a)   (b)

Figure 3.22: Magnetization and magnetic field distribution of a surface wave mode (MSSW) for $k_z = 0$ [13]: (a) transverse magnetization and (b) transverse magnetic fields.

At the limit of very large $k_{xi} = j\alpha_i$, the $f_i(x)$ eigenfunction can be approximated by exponential functions. First, we should note that $f_i(x)$ in (3.149) represents both modes with the same k_y. The first term clings onto the top surface $x = +d/2$, propagating in the negative y-direction as $e^{+jk_y y}$, while the second term clings onto the bottom surface $x = -d/2$, propagating in the positive y-direction as $e^{-jk_y y}$. Note that this phenomenon is also described in [40, 41] but with opposite signs for k_y since they assume an $e^{-j\omega t}$ time dependence instead of an $e^{+j\omega t}$ used herein. Hence, the two modes can be written as:

Mode clinging onto the top surface $x = +d/2$ and propagating toward negative $\hat{y}$:

$$\Psi(x) = \Psi_{0e} e^{jk_y y} e^{-jk_z} \frac{\sinh[\alpha_i(x+d/2)]}{\sinh(\alpha_i d)} \bigg|_{\alpha_i d/2 >> 1} \approx \Psi_{0e} e^{jk_y y} e^{-jk_z} \frac{e^{\alpha_i(x+d/2)}}{e^{\alpha_i d}}$$

$$\approx \Psi_{0e} e^{jk_y y} e^{-jk_z} e^{\alpha_i(x-d/2)} \quad \text{for} \quad |x| \leq d/2. \tag{3.164a}$$

Mode clinging onto the bottom surface $x = -d/2$ and propagating toward the positive $\hat{y}$:

$$\Psi(x) = \Psi'_{0e} e^{-jk_y y} e^{-jk_z} \frac{\sinh\left[\alpha_i\left(x - d/2\right)\right]}{\sinh\left(\alpha_i d\right)} \bigg|_{\alpha_i d/2 \gg 1} \approx \Psi'_{0e} e^{-jk_y y} e^{-jk_z} \frac{e^{-\alpha_i(x-d/2)}}{e^{\alpha_i d}}$$

$$\approx \Psi'_{0e} e^{-jk_y y} e^{-jk_z} e^{-\alpha_i(x+d/2)} \quad \text{for} \quad |x| \leq d/2. \tag{3.164b}$$

A graphic representation of the above case is shown in Figure 3.23. It is important to note that *non-reciprocal* MSSW modes follow the right-hand rule, circulating around the positive z-direction, along which the DC bias and saturation magnetization are assumed; in other words, align your right thumb along the static DC biasing field and turn your hand clockwise to show the direction of the circulating MSSW field. Another important feature of these MSSW modes is their circular (or elliptic) polarization [41].

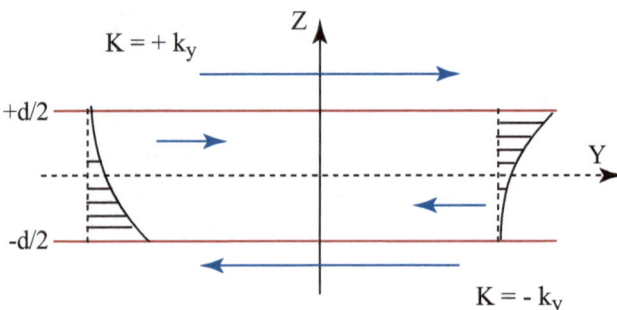

Figure 3.23: Graphic representation of the two non-reciprocal MSSW modes for the limiting case when $\alpha_i d \gg 1$ [41].

3.23.3   POLARIZATION OF MAGNETOSTATIC WAVES

It is likely that circularly or elliptically polarized waves exist in ferrites due to the 90° phase difference between the diagonal and off-diagonal elements of the permeability tensor. In turn, this circular polarization appears directly in the RF magnetic field components that are transverse to DC magnetization. In the present case, this would occur in h_x and h_y since the DC bias is along the z-axis. Recall that the RF magnetic field $\bar{h}$ is expressed in terms of the magnetic potential ψ_i as $\bar{h} = -\nabla\Psi_i$. Now, a question appears about the modes for which the differentiation causes a 90° (namely, an additional j) phase difference between h_x and h_y.

Concerning volume magnetostatic modes (MSBVW in longitudinal bias), the differentiation shifts both h_x and h_y by 90°. This happens because the former has an exponential $e^{jk_y y}$ dependence, while the latter has a sinusoidal one, which is also equivalent to $e^{\pm jk_x x}$. In contrast, for surface modes (MSSW), the x-dependence becomes hyperbolic or $e^{\pm \alpha_i x}$. Thus, it is obvious that the differentiation of Ψ will shift h_y by $\pm j = e^{\pm j90°}$ but not h_x, causing in general elliptical polarization. Hence, MSBVW will be linearly polarized, while MSSW modes will be elliptically (or circularly) polarized. With this opportunity, an examination of expression (3.91) for transverse magnetization shows that magnetostatic forward volume waves (MSFVW) are also linearly polarized.

3.23.4 POLARIZATION OF MSSW MODES

For an analytical examination of the polarization of MSSW modes, let us rewrite the magnetic potential expression in the ferrite slab region with the help of (3.113) and (3.149) as:

$$
\Psi_i = f_i(x)g(y)h(z) = \frac{1}{\sinh(\alpha_i d)}
$$
$$
\left\{ \Psi_{0e} \sinh\left[\alpha_i \left(x + \frac{d}{2}\right)\right] - \Psi'_{0e} \sinh\left[\alpha_i \left(x - \frac{d}{2}\right)\right] \right\} e^{jk_y y} e^{-jk_z z}. \tag{3.165}
$$

In turn, the transverse RF magnetic-field components read as follows:

$$
h_x = \frac{-\partial \Psi_i}{\partial x} = \frac{-\alpha_i}{\sinh(\alpha_i d)}
$$
$$
\left\{ \Psi_{0e} \cosh\left[\alpha_i \left(x + \frac{d}{2}\right)\right] - \Psi'_{0e} \cosh\left[\alpha_i \left(x - \frac{d}{2}\right)\right] \right\} e^{jk_y y} e^{-jk_z z} \tag{3.166a}
$$
$$
h_y = \frac{-\partial \Psi_i}{\partial y} = \frac{-1}{\sinh(\alpha_i d)}
$$
$$
\left\{ \Psi_{0e} \sinh\left[\alpha_i \left(x + \frac{d}{2}\right)\right] - \Psi'_{0e} \sinh\left[\alpha_i \left(x - \frac{d}{2}\right)\right] \right\} (jk_y) e^{jk_y y} e^{-jk_z z}. \tag{3.166b}
$$

Remember that the above expressions correspond to the complex phasors usually employed for the analysis of time-harmonic fields. Their formulas in the time domain are:

$$
h_x(t) = \text{Re}\left(h_x e^{j\omega t}\right) = -f'_i(x) \cos\left(k_y y - k_z z + \omega t\right) \tag{3.167a}
$$
$$
h_y(t) = \text{Re}\left(h_y e^{j\omega t}\right) = +f_i(x) k_y \sin\left(k_y y - k_z z + \omega t\right). \tag{3.167b}
$$

The angle of the transverse magnetic field $\bar{h}_t = h_x \hat{x} + h_y \hat{y}$ with respect to the x-axis, (Figure 3.24), can be defined as:

$$
\tan \theta = \frac{h_y}{h_x} \quad \Leftrightarrow \quad \theta = \tan^{-1}\left\{ (-k_y) \frac{f_i(x)}{f'_i(x)} \tan\left(k_y y - k_z z + \omega t\right) \right\}. \tag{3.168}
$$

Recall at this point the basic definition of rotating, e.g., Pozar [3], for circular polarization:

$$
\bar{H}_t = H_{01} \hat{x} - j H_{02} \hat{y}. \tag{3.169a}
$$

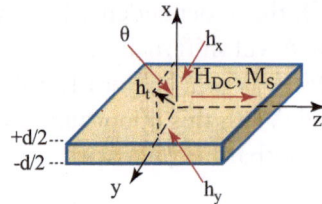

Figure 3.24: Definition of the transverse RF magnetic field $\bar{h}_t$ and its angular dependence.

When H_{01} and H_{02} are real values, polarization is right-hand elliptical (RHEP) while the expression

$$\bar{H}_t = H_{01}\hat{x} + jH_{02}\hat{y},\tag{3.169b}$$

yields a left-hand elliptical polarization (LHEP).

Hence, (3.168) shows that polarization is generally elliptical, and the direction of rotation depends on the sign of k_y, which is shown above to be different at the top $x = +d/2$ and at the bottom $x = -d/2$ of the slab.

For a more convenient interpretation, let us examine the polarization on these surfaces in the limiting case in which $\alpha_i d \gg 1$ by exploiting expressions (3.164).

From (3.164a), on the top surface $x = +d/2$ we have

$$h_x \approx -\Psi_{0e} \cdot \alpha_i \cdot e^{\alpha_i(x-d/2)} e^{jk_y y} e^{-jk_z z}\tag{3.170a}$$

$$h_y \approx -\Psi_{0e}\left(jk_y\right) e^{\alpha_i(x-d/2)} e^{jk_y y} e^{-jk_z z}.\tag{3.170b}$$

Taking the real values in the time domain

$$h_x(t) = -\Psi_{0e}\alpha_i \, e^{\alpha_i(x-d/2)} \cos\left(k_y y - k_z z + \omega t\right)\tag{3.171a}$$

$$h_y(t) = +\Psi_{0e}k_y \, e^{\alpha_i(x-d/2)} \sin\left(k_y y - k_z z + \omega t\right),\tag{3.171b}$$

the angle of rotation becomes

$$\theta = \tan^{-1}\left[-\frac{k_y}{\alpha_i} \tan\left(k_y y - k_z z + \omega t\right)\right].\tag{3.172a}$$

Both k_y and α_i are real and positive and for the case in which $k_z \approx 0$ expression (3.151) gives $k_y = \alpha_i$, which yields

$$\theta \approx -\left(k_y y + \omega t\right).\tag{3.172b}$$

Observing (3.171), the amplitudes of h_x and h_y become equal (circular orbit), and from (3.172) we can see that the magnetic field h_t rotates clockwise or in a left-hand direction with respect to the positive z-axis. Therefore, this is a left-hand circularly polarized wave (LHCP) propagating in the negative y-direction. Note that at the point $y = 0$ is clearly $\theta = -\omega t$; that is, this is an anti-Larmor CP wave. Likewise, from (3.164b) on the bottom surface $x = -d/2$ we have

$$h_x \approx -\Psi'_{0e}\left(-\alpha_i\right)\, e^{-\alpha_i(x+d/2)} e^{-jk_y y} e^{-jk_z z} \tag{3.173a}$$

$$h_y \approx -\Psi'_{0e}\left(-jk_y\right)\, e^{-\alpha_i(x+d/2)} e^{-jk_y y} e^{-jk_z z}, \tag{3.173b}$$

and

$$h_x(t) = \Psi'_{0e}\, \alpha_i\, e^{-\alpha_i(x+d/2)} \cos\left(-k_y y - k_z z + \omega t\right) \tag{3.174a}$$

$$h_y(t) = -\Psi'_{0e}\, k_y\, e^{-\alpha_i(x+d/2)} \sin\left(-k_y y - k_z z + \omega t\right), \tag{3.174b}$$

yielding

$$\theta \approx \tan^{-1}\left[-\frac{k_y}{\alpha_i}\tan\left(-k_y y - k_z z + \omega t\right)\right]. \tag{3.175a}$$

Note that the opposite sign as k_y is already considered. Therefore, k_y and α_i are both positive. Once again, for $k_z \approx 0$ we have $k_y \approx \alpha_i$. Thus,

$$\theta \approx -\left(-k_y y + \omega t\right). \tag{3.175b}$$

Again, this is an anti-Larmor LHCP wave propagating in the positive y-direction. Considering the above two cases, this wave rotates clockwise, propagating on the surface and traveling around the slab in the y-direction. Moreover, according to [41] and the original references given therein, these waves are *non-resonant*.

3.23.5   GRAPHIC REPRESENTATION OF MAGNETOSTATIC WAVES

To sum up this section, it is useful to give a graphic representation of the three categories of magnetostatic waves for the most common cases [16].

The magnetostatic forward volume wave (MSFVW) occurs when the slab or film is transversely magnetized (z-axis). The field propagates in both longitudinal ($\hat{x}, \hat{y}$) directions and its energy is spread all over the slab volume, following either a symmetric or anti-symmetric distribution in the transverse $\hat{z}$-direction. A graphic representation of a wave propagating in the x-direction is shown in Figure 3.25.

Magnetostatic backward volume waves (MSBVW) occur when the slab or film is magnetized longitudinally ($\hat{z}$-axis), with their energy distributed all over the slab volume. The wave is generally assumed to propagate in both longitudinal ($\hat{y}, \hat{z}$) directions, but pure propagation

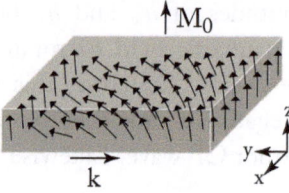

Figure 3.25: Graphic representation of an MSFVW [16].

transverse to the DC magnetization ($k_z = (0)$ is not possible for volume modes. In contrast, pure propagation along the DC magnetization ($k_y = (0)$ is the common case and an indicative example is presented in Figure 3.26. Their special characteristic is the negative group velocity; that is, the direction of energy flow is opposite to the direction of propagation (direction of wavenumber and phase velocity).

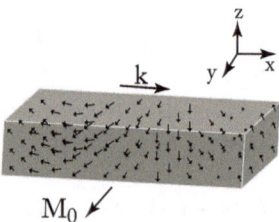

Figure 3.26: Graphic representation of an MSBVW [16].

The magnetostatic surface wave occurs when the slab is longitudinally magnetized ($\hat{z}$-direction) but at frequencies outside the spin-wave manifold (volume waves fall within it). These waves primarily propagate transversely to the DC bias direction ($\hat{y}$-direction), as shown in the indicative example of Figure 3.27. Their energy clings to either the top or the bottom surface of the slab, following an anti-Larmor circular polarization and traveling around the slab.

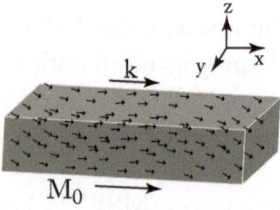

Figure 3.27: Graphic representation of an MSSW [16].

Before moving on to practical, useful, multilayered structures, like dielectric ferrites or ferroelectric ferrite films, it is worth studying magnetostatic (or spin) waves in some geometries of circular symmetry (e.g., disks and rods). Another important issue is the damping of spin and magnetostatic waves, since losses often reveal some hidden physical phenomena while always better representing practical applications.

3.23.6   MAGNETOSTATIC MODES IN AN INFINITE CIRCULAR DISK-PERPENDICULAR MAGNETIZATION

To simplify the analysis, let us start with a disk of infinite radius, which was originally studied by Sparks [21, 42]. This is the same geometry as studied in a past section. The basic difference here is the employment of circular-cylindrical coordinates ($\hat{\rho}$, $\hat{\phi}$, $\hat{z}$). Hence, Walker's Eq. (3.76) should be appropriately expressed. Recall that the permeability tensor $[\mu]$ as well as the susceptibility $\mathcal{X}$ retains the same form for the ($\hat{\rho}$, $\hat{\phi}$, $\hat{z}$) and ($\hat{x}$, $\hat{y}$, $\hat{z}$) systems. The former is given in (2.21). The wave equation in its general form is given in (3.76). To expand it for the ferrite disk, let us write the grad operator $\nabla\Psi^i$ in cylindrical coordinates. Then multiply this by the permeability tensor (3.73) as $[\mu]\nabla\Psi^i$. One may observe then the occurrence of two unexpected terms containing $\pm jk_1$, which cancel each other out when we take the divergence $\nabla \cdot ([\mu]\nabla\Psi^i)$. The resulting wave equations for both the ferrite disk and the air region agree with those of Damon and Van De Vaart [37], and read as follows:

Ferrite disk:
$$(1 + \mathcal{X})\left\{\frac{1}{\rho}\frac{\partial}{\partial\rho}\left(\rho\frac{\partial\Psi^i}{\partial\rho}\right) + \frac{1}{\rho^2}\frac{\partial^2\Psi^i}{\partial\varphi^2}\right\} + \frac{\partial\Psi^i}{\partial z^2} = 0 \qquad (3.176a)$$

Air outside the sample:
$$\frac{1}{\rho}\frac{\partial}{\partial\rho}\left(\rho\frac{\partial\Psi^e}{\partial\rho}\right) + \frac{1}{\rho^2}\frac{\partial^2\Psi^e}{\partial\varphi^2} + \frac{\partial\Psi^e}{\partial z^2} = 0 \qquad (3.176b)$$

Recall that the demagnetization factors are the same as those noted in Chapter 2 as $N_z = 1$ and $N_x = N_y = 0 \leftrightarrow N_\rho = N_\varphi = 0$ or the internal DC magnetic field is $H_i = H_0 - M_s$, oriented along the normal z-axis. In addition, the same arguments concerning the boundary condition along the z-axis at the Ferrite-air interface $z = \pm d/2$ should be imposed. For a general solution of (3.176), let us apply the separation of variables as $\Psi(\rho, \varphi, z) = f(\rho)g(\varphi)h(z)$. First, the field continuity conditions at the ferrite-air interface call for the same dependence in ρ and φ coordinates, or $g^e(\varphi) = g^i(\varphi) = g(\varphi)$ and $f^e(\rho) = f^i(\rho) = f(\rho)$, namely:

$$\Psi^e(\rho, \varphi, z) = f(\rho)g(\varphi)h^e(z) \qquad (3.177a)$$

$$\Psi^i(\rho, \varphi, z) = f(\rho)g(\varphi)h^i(z). \qquad (3.177b)$$

It is worth elaborating on the ferrite case, applying the separation of variables in (3.176a) to yield

$$\rho \frac{\partial}{\partial \rho} \left(\rho \frac{\partial f(\rho)}{\partial \rho} \right) + \left\{ -\frac{k_{zi}^2}{(1+\mathcal{X})} \rho^2 - n^2 \right\} f(\rho) = 0. \tag{3.178}$$

In the φ-direction, the eigenfunction should be of the form $e^{-jn\varphi}$ or $e^{jn\varphi}$, where $n = 0, 1, 2, \ldots$ due to the 2π periodicity in angle-φ. Moreover, in the z-direction, a standing wave with wavenumber k_{zi} should be assumed inside the ferrite:

$$h^i(z) = A_z \cos(k_{zi}z) + B_z \sin(k_{zi}z). \tag{3.179}$$

In order to satisfy the radiation condition at infinity, an exponential decay must be considered in the air region as $e^{-\alpha z}$ or $k_{ze} = -j\alpha$. Equation (3.176) can be adapted for the air region by just setting $\mathcal{X} \to 0$ and $k_{zi} \to k_{ze}$. The radial wavenumber k_ρ can then be defined as:

$$k_\rho^2 = -k_{ze}^2 = -(-j\alpha)^2 = \alpha^2 \quad \Leftrightarrow \quad k_\rho = \alpha. \tag{3.180a}$$

To get a volume magnetostatic wave, k_{zi} should be real. Furthermore, for the field to propagate in the ρ-direction, the corresponding wavenumber (k_ρ) should also be real. Observing (3.178) one may write

$$k_\rho^2 = -k_{zi}^2/(1+\mathcal{X}) \quad \Leftrightarrow \quad k_\rho = k_{zi}/\sqrt{-(1+\mathcal{X})}. \tag{3.180b}$$

Asking for k_ρ to be real is equivalent to the expected search for volume magnetostatic waves in the frequency range defined by $(1 + \mathcal{X}) < 0$ and given in Eq. (3.133b). In turn, (3.178) reads as follows:

$$\rho^2 \frac{d^2 f(\rho)}{d\rho^2} + \rho \frac{df(\rho)}{d\rho} + \left[k_\rho^2 \cdot \rho^2 - n^2 \right] f(\rho) = 0. \tag{3.181}$$

This is an ordinary Bessel differential equation. When an infinitely extending slab is considered, the appropriate general solutions are Hankel's functions $H_n^{(2)}(k_\rho\rho)$ of the second kind, which represent waves propagating in the positive $\hat{\rho}$ direction for the assumed $e^{j\omega t}$ temporal dependence. It should be noticed that $H_n^{(1)}(k_\rho\rho)$ represents waves propagating in the negative $\hat{\rho}$ direction. Since the final goal is to study finite disks, it is preferable to consider standing-wave solutions in terms of Bessel's $J_n(k_\rho\rho)$ or Newman's functions $Y_n(k_\rho\rho)$. Furthermore, $Y_n(k_\rho\rho)$ tends to infinity at $\rho \to 0$. Thus, only $J_n(k_\rho\rho)$ can be adopted.

Furthermore, the corresponding analysis in Cartesian coordinates revealed that the normally biased infinite film supports both symmetric (or even) and anti-symmetric (or odd) modes, proportional to $\cos(k_{zi}z)$ and $\sin(k_{zi}z)$, respectively. However, both modes can be combined into a single expression by exploiting a simple trigonometric identity:

$$\cos\left(z - \frac{n\pi}{2}\right) = \begin{cases} (-1)^{n/2} \cos(z) & \text{even modes} \quad n = 0, 2, 4, \ldots \\ (-1)^{(n-1)/2} \sin(z) & \text{odd modes} \quad n = 1, 3, 5, \ldots. \end{cases} \tag{3.182}$$

Concerning the azimuthal dependence, only the $e^{jn\varphi}$ is allowed for reasons explained below.

In view of the above, a general solution, which is also in agreement with that of Sparks [21], can be obtained. Thus,

Ferrite disk: $\Psi_i = \Psi_{0i} J_n(k_\rho \rho) e^{jn\varphi} \cdot \cos\left(k_{zi}z - \frac{n\pi}{2}\right)$ for $-\frac{d}{2} \leq z \leq \frac{d}{2}$ (3.183a)

Air region: $\Psi_e = \begin{cases} \Psi_{0e} J_n(k_\rho \rho) e^{jn\varphi} \cdot e^{-\alpha(z-d/2)} & \text{for} \quad z \geq d/2 \\ \Psi'_{0e} J_n(k_\rho \rho) e^{jn\varphi} \cdot e^{+\alpha(z+d/2)} & \text{for} \quad z \leq -d/2. \end{cases}$ (3.183b)

Both continuity conditions of the tangential magnetic field ($h_\rho = -\partial\Psi/\partial\rho$ and $h = -(1/\rho)\partial\Psi/\partial\varphi$) are reduced to the same expressions:

At $z = +d/2$: $\Psi_{0i} \cos(k_{zi}d/2 - n\pi/2) = \Psi_{0e}$ (3.184a)

At $z = -d/2$: $\Psi_{0i} \cos(k_{zi}d/2 + n\pi/2) = \Psi'_{0e}.$ (3.184b)

On the other hand, the continuity of the normal flux density $b_n = b_z = -\mu_0 \partial\Psi/\partial z$ yields

At $z = +d/2$: $-k_{zi}\Psi_{0i} \sin(k_{zi}d/2 - n\pi/2) = -\alpha\Psi_{0e}$ (3.185a)

At $z = -d/2$: $+k_{zi}\Psi_{0i} \sin(k_{zi}d/2 + n\pi/2) = +\alpha\Psi'_{0e}.$ (3.185b)

By expanding either the cosinus in (3.184) or the sinus in (3.185), one may easily show that $\Psi'_{0e} = \Psi_{0e}$ for even modes $n = 0, 2, 4, \ldots$; namely, even modes are symmetric with respect to the slab midplane $z = 0$. Likewise, $\Psi'_{0e} = -\Psi_{0e}$ for odd modes $n = 1, 3, 5, \ldots$, which are anti-symmetric with respect to $z = 0$. In turn, dividing (3.185) by (3.184) yields the expected characteristic equation already given in (3.107a):

MSFVW waves: $\tan\left(k_{zi}\frac{d}{2} - \frac{n\pi}{2}\right) = \frac{\alpha}{k_{zi}} = \frac{k_\rho}{k_{zi}}.$ (3.186)

The remaining analysis given in this section is essentially the same and leads to the same mode characteristics. What would be interesting here is to consider a ferrite disk of finite radius. However, in this case, the general solution must account for the additional boundary conditions at the ferrite disk perimeter $\rho = R$. The key to this analysis is zero magnetization in the assumed air region $\rho > R$.

3.23.7   FINITE FERRITE DISKS

Consider a transversely magnetized circular ferrite disk with radius $\rho = R$. The general solution for the scalar magnetic potential Ψ inside the disk is given again by (3.183); that is, it remains the same as that for an infinitely extending disk. However, magnetization outside the disk in the region $\rho > R$ should be zero. This requirement provides the key boundary conditions at $\rho = R$ which accounts for the disk's finite extent. Recall that magnetization $\bar{m}$ is related to the magnetic field $\bar{h}$ through the susceptibility tensor $[\mathcal{X}]$ by:

$$\bar{m} = [\mathcal{X}]\bar{h}. \tag{3.187}$$

The $[\mathcal{X}]$ elements are given in (3.70). In turn, $\bar{h}$ is related to the magnetic potential Ψ, which must be expanded now in cylindrical coordinates as follows:

$$\bar{h} = -\nabla\Psi^i = -\left(\hat{\rho}\frac{\partial\Psi^i}{\partial\rho} + \hat{\varphi}\frac{1}{\rho}\frac{\partial\Psi^i}{\partial\varphi} + \hat{z}\frac{\partial\Psi^i}{\partial z}\right). \tag{3.188}$$

It has already been explained in the previous sections that $\bar{m}$ represents the microwave magnetization, which under the small signal approximation has a negligible component in the DC biasing direction, that is, $m_z \approx 0$. Substituting (3.188) into (3.187) yields

$$\bar{m} = -[\mathcal{X}]\left(\nabla\Psi^i\right) \quad \text{for} \quad \rho \le R \quad \text{and} \quad -d/2 \le z \le d/2. \tag{3.189}$$

By exploiting the general solution for Ψ^i given in (3.183a) and using the susceptibility definition, the m_ρ and m_φ components read as follows:

$$m_\rho = -\mathcal{X}h_\rho - jk_1h_\varphi = -\mathcal{X}\frac{\partial\Psi^i}{\partial\rho} - jk_1\frac{1}{\rho}\frac{\partial\Psi^i}{\partial\varphi} \tag{3.190a}$$

$$m_\varphi = jk_1h_\rho - \mathcal{X}h_\varphi = jk_1\frac{\partial\Psi^i}{\partial\rho} - \mathcal{X}\frac{1}{\rho}\frac{\partial\Psi^i}{\partial\varphi}. \tag{3.190b}$$

Carrying out the differentiations, the above expressions yield

$$m_\rho = \psi_{0i}e^{jn\varphi}\cos\left(k_{zi}z - n\frac{\pi}{2}\right)\left\{-\mathcal{X}\cdot k_\rho\cdot J_n'\left(k_\rho\rho\right) + k_1\cdot\frac{n}{\rho}\cdot J_n\left(k_\rho\rho\right)\right\} \tag{3.191a}$$

$$m_\varphi = j\psi_{0i}e^{jn\varphi}\cos\left(k_{zi}z - n\frac{\pi}{2}\right)\left\{k_1\cdot k_\rho\cdot J_n'\left(k_\rho\rho\right) - \mathcal{X}\cdot\frac{n}{\rho}\cdot J_n\left(k_\rho\rho\right)\right\}. \tag{3.191b}$$

From (3.191) it can be observed that asking for both m_ρ and m_φ to vanish at the disk edge $\rho = R$ yields two inconsistent equations, at least at first glance. This requirement is also known as "spin pinning" at the ferrite surface. According to Maksymowitz [43], for example, the inconsistencies may arise due to localized non-uniformities around the disk edges. Concerning the demagnetization factors of the infinite disk, it was assumed that $N_z = 1$ and $N_x = N_y = 0$, but this is violated around the edge, where $\rho = R$. However, this non-uniformity is localized and disappears at a small distance (inward) from the edge. Nevertheless, pinning the spins at the disk edge requires that:

$$m_\rho(\rho = R) = 0 \leftrightarrow -k_\rho \left\{ \mathcal{X} \cdot J_n'\left(k_\rho R\right) - k_1 \cdot \frac{n}{k_\rho R} \cdot J_n\left(k_\rho R\right) \right\} = 0 \qquad (3.192a)$$

$$m_\varphi(\rho = R) = 0 \leftrightarrow k_\rho \left\{ k_1 \cdot J_n'\left(k_\rho R\right) - \mathcal{X} \cdot \frac{n}{k_\rho R} \cdot J_n\left(k_\rho R\right) \right\} = 0. \qquad (3.192b)$$

For these equations to have a non-trivial solution, $\mathcal{X}$ and k_1 must be approximately equal, $\mathcal{X} \approx k_1$, or equivalently, the circular frequency (ω) must be around ω_i, that is, $\omega \approx \omega_i$. These may also be true in the limit $k_\rho R \gg 1$, which was first stated in [21]. In view of this and the formula for the Bessel functions derivatives, e.g., Abramouvitz and Stegun [39],

$$J_n'(z) = -J_{n+1}(z) + \frac{n}{z} J_n(z) = J_{n-1}(z) - \frac{n}{z} J_n(z). \qquad (3.193)$$

Both (3.192) conditions are reduced to:

$$J_{n-1}\left(k_\rho R\right) = 0. \qquad (3.194)$$

Likewise, the magnetization components can be written as:

$$m_\rho = \psi_{0i} \left(\mathcal{X} \cdot k_\rho\right) J_{n+1}\left(k_\rho \rho\right) \cdot e^{jn\varphi} \cdot \cos\left(k_{zi} z - n\frac{\pi}{2}\right) \qquad (3.195a)$$

$$m_\varphi = \psi_{0i} \left(-j\mathcal{X} \cdot k_\rho\right) J_{n+1}\left(k_\rho \rho\right) \cdot e^{jn\varphi} \cdot \cos\left(k_{zi} z - n\frac{\pi}{2}\right). \qquad (3.195b)$$

Besides the above, the boundary conditions at $z = \pm d/2$ still require the validity of (3.186), but now k_ρ should satisfy (3.194). The consequence of these requirements is that the radial wavenumber k_ρ takes only discrete values for a finite disk instead of being a continuous variable in the infinite film case. By symbolizing the mth root of the Bessel function J_{n+1} with $P_{n+1,m}$, k_ρ is obtained as

$$J_{n+1}\left(P_{n+1,m}\right) = 0 \quad \leftrightarrow \quad k_\rho = P_{n+1,m}/R. \qquad (3.196)$$

These roots are given in Table 3.1 below, e.g., Pozar [3].

A variational calculation applied on (3.192) yields a resonance frequency for the dominant $(n + 1 = 0)$ mode [42], as:

$$\omega_d \approx \frac{1}{2}\omega_m \cdot \left\{ 1 - \frac{1 - e^{k_\rho/d}}{k_\rho d} \right\}. \qquad (3.197)$$

This expression is extracted without any pinning at the interfaces $z = \pm d/2$.

Table 3.1: Roots of $J_{n+1}(P_{n+1,m}) = 0$

n + 1	m = 1	m = 2	m = 3
0	2.405	5.520	8.654
1	3.832	7.016	10.174
2	5.135	8.417	11.620

3.23.8   FINITE CYLINDRICALLY SYMMETRIC SAMPLES–RODS

The analysis of the finite ferrite disk presented in the previous section was based on the general solution and the boundary conditions for magnetization. Since some mathematical difficulties are observed when the solution is restricted to the ferrite domain, it is interesting to go through an approach based on the scalar magnetic potential. Even though this approach seems more complicated at first glance, it is straightforward and describes the field in the unbounded domain. Fletcher and Kittel [44] first presented this approach for the analysis of an infinite ferrite rod. Also, Joseph and Schlomann [45], studied the magnetostatic modes in long axially magnetized cylinders. For a finite disk [46], the general solution in the ferrite region is still that of Eq. (3.183). In contrast, the ferrite rod is infinitely extended in the z-direction along which a propagating solution should be considered as $\propto e^{-jk_z z} = e^{-j\beta z}$:

$$\text{Ferrite rod:} \qquad \psi^i = \psi_{0i} J_n(k_\rho\rho) e^{jn\varphi} e^{-j\beta z} \quad \text{for} \quad \rho \leq R. \qquad (3.198)$$

Outside the ferrite ($\rho > R$) a radial eigenfunction satisfying the Sommerfield radiation condition at infinity should be employed. This can be either a McDonald $K_n(k_{\rho 0}\rho)$ or a Hankel function of the second kind $H_n^{(2)}(k_{\rho 0}\rho)$, in which $k_{\rho 0}$ is the radial wavenumber in the unbounded region. This selection is justified by the large argument asymptotic expansion of these functions [39]:

$$K_n(z)|_{|z|\to\infty} \approx \sqrt{\pi/2z} \cdot e^{-z} \qquad (3.199a)$$

$$H_n^{(2)}(z)\Big|_{|z|\to\infty} \approx \sqrt{2/\pi z} \cdot e^{+j(n\pi/2+\pi/4)} e^{-jz}. \qquad (3.199b)$$

Recall also that for the time dependence $e^{+j\omega t}$ considered herein, the Hankel function of the second kind $H_n^{(2)}(k_{\rho 0}\rho) = J_n(k_{\rho 0}\rho) - jY_n(k_{\rho 0}\rho)$ represents an outward propagating wave toward positive $\hat{\rho}$, while its first kind counterpart $H_n^{(1)}(k_{\rho 0}\rho) = J_n(k_{\rho 0}\rho) + jY_n(k_{\rho 0}\rho)$ represents inward propagating waves. It is also obvious from (3.199a) that $K_n(k_{\rho 0}\rho)$ solely represents surface waves exponentially attenuated in the positive $\hat{\rho}$-direction and the same is true for the $H_n^{(2)}$ with an imaginary argument. In addition, note that $k_n(z)$ is also a solution of the modified Bessel equation [39]:

$$z^2 \frac{d^2 w}{dz^2} + z \frac{dw}{dz} - (z^2 + n^2) w = 0. \qquad (3.200)$$

Comparing (3.200) to (3.181), one may notice that $k_{\rho 0}^2$ corresponds to $-k_\rho^2$ and by setting $\mathcal{X} = 0$ to denote air instead of ferrite (3.178) yields:

$$- k_{\rho 0}^2 = -k_{z0}^2 \quad \text{or} \quad k_{\rho 0}^2 = k_{z0}^2 = k_{zi}^2 \leftrightarrow k_{\rho 0} = k_{zi}. \tag{3.201}$$

Moreover, the continuity conditions must hold for an arbitrary z-coordinate, requiring the same z-dependence inside and outside the ferrite $k_{z0} = k_{zi}$. In view of the above, the general solutions outside the sample can be formulated as follows:

Ferrite rod: $\quad \psi_{eF} = \psi_{0F} K_n \left(k_{\rho 0} \rho \right) e^{jn\varphi} \cdot e^{-j\beta z} \quad \rho \geq R \tag{3.202a}$

Ferrite disk: $\quad \psi_{eF} = \psi_{0F} K_n \left(k_{\rho 0} \rho \right) e^{jn\varphi} \cos \left(k_{zi} z - \dfrac{n\pi}{2} \right) \rho \geq R, -\dfrac{d}{2} \leq z \leq \dfrac{d}{2}. \tag{3.202b}$

The continuity conditions of the tangential magnetic field at $\rho = R : h_z = -\partial\psi/\partial z$ and $h_\varphi = -(1/\rho)\partial\psi/\partial z$ are both reduced to a single expression for either rod or disk samples:

$$\psi_{0i} J_n \left(k_\rho R \right) = \psi_{0F} K_n \left(k_{\rho 0} R \right). \tag{3.203}$$

Likewise, the continuity of the normal flux density

$$b_n = b_\rho = \begin{cases} -\mu_0(1 + \mathcal{X})\dfrac{\partial\psi}{\partial\rho} - j\mu_0 k_1 \dfrac{1}{\rho}\dfrac{\partial\psi}{\partial\varphi} & \rho \leq R \\ -\mu_0 \dfrac{\partial\psi}{\partial\rho} & \rho \geq R. \end{cases} \tag{3.204}$$

at $\rho = R$ yields

$$\psi_{0i} \left[(1 + \mathcal{X}) \cdot k_\rho \cdot J_n' \left(k_\rho R \right) - k_1 \dfrac{n}{R} J_n \left(k_\rho R \right) \right] = \psi_{0F} \cdot k_{\rho 0} \cdot K_n' \left(k_{\rho 0} R \right). \tag{3.205}$$

By dividing (3.205) by (3.203), the characteristic equation is obtained as follows:

$$(1 + \mathcal{X}) \cdot k_\rho \cdot \dfrac{J_n' \left(k_\rho R \right)}{J_n \left(k_\rho R \right)} - \dfrac{k_1 \cdot n}{R} = k_{\rho 0} \cdot \dfrac{K_n' \left(k_{\rho 0} R \right)}{K_n \left(k_{\rho 0} R \right)}. \tag{3.206a}$$

Remember that k_ρ and $k_{\rho 0}$ are both related to k_{zi} through (3.180b) and (3.201) respectively, which when substituted above, (3.206a) reads as follows:

$$\sqrt{-(1 + \mathcal{X})} \cdot \dfrac{J_n' \left(k_{zi}R / \sqrt{-(1 + \mathcal{X})} \right)}{J_n \left(k_{zi}R / \sqrt{-(1 + \mathcal{X})} \right)} + \dfrac{K_n' \left(k_{zi}R \right)}{K_n \left(k_{zi}R \right)} + \dfrac{k_1 n}{R \cdot k_{zi}} = 0. \tag{3.206b}$$

The above equation is in accordance with those of [45] and [46]. The only difference is that the azimuthal dependence $e^{-jn\varphi}$ is assumed therein. It is also interesting to examine the short wavelength limit [44], where $k_\rho \rho \gg 1$. In this case, the McDonald function may be approximated through (3.199a). Since it is a monotonically decreasing and not an oscillating function, its derivative is approximated by [39] as:

$$K_n'(z)\big|_{|z| \to \infty} \approx -\sqrt{\frac{\pi}{2z}}\, e^{-z} \approx -K_n(z). \qquad (3.207)$$

By exploiting (3.207), the limiting form of (3.206b) reads as follows:

$$\sqrt{-(1+\mathcal{X})} \cdot \frac{J_n'\left(k_{zi}R/\sqrt{-(1+\mathcal{X})}\right)}{J_n\left(k_{zi}R/\sqrt{-(1+\mathcal{X})}\right)} - 1 + \frac{k_1 n}{k_{zi}R} \approx 0 \quad \text{for} \quad k_{zi}R \gg 1. \qquad (3.208)$$

Recall that volume modes are only possible in the range set by Eq. (3.133b), in which $(1+\mathcal{X}) < 0$. The lower-order modes $n = 1$ and solutions $k_{zi}R \gg 1$ are obtained when [44]:

$$\mathcal{X} \approx k_1 \approx -1/(2\mathcal{E}) \quad \text{where} \quad \mathcal{E} = (\omega - \omega_0)/\omega_m \ll 1. \qquad (3.209)$$

By letting $X_{1m} = k_{zi}R/\sqrt{-(1+\mathcal{X})} \approx \sqrt{2\mathcal{E}} \cdot k_{zi}R$, (3.209) yields:

$$J_1'(X_{1m})/J_1(X_{1m}) \approx -1/X_{1m} \qquad (3.210a)$$

which is also reduced to:

$$J_1'(X_{1m}) \approx 0 \quad \leftrightarrow \quad J_0(X_{1m}) \approx 0. \qquad (3.210b)$$

The latter is in turn identical to (3.196) with the three lower roots $X_{1m} = 2.405, 5.520, 8.654$ for $m = 1, 2,$ and 3, respectively. The lowest eigenfrequency for $X_{11} = 2.405$ is

$$\omega_1 \approx \omega_0 + \mathcal{E} \cdot \omega_m = \omega_0 + \frac{1}{2}\left(\frac{2.405}{k_{zi}R}\right)^2 \cdot \omega_m \qquad (3.211a)$$

where $\mathcal{E}$ is substituted by $\mathcal{E} \approx (X_{1m}/k_{zi}R)^2$.

The exchange energy may be considered in (3.211a) by adding the term $Dk^2 = Dk_{zi}^2$, which results in [44],

$$\omega_1 \approx \omega_0 + \frac{1}{2}\left(\frac{2.405}{k_{zi}R}\right)^2 \cdot \omega_m + D \cdot k_{zi}^2. \qquad (3.211b)$$

The group velocity for the first mode is then calculated through (3.211a):

$$V_{g1} \approx \frac{\partial \omega_1}{\partial k} = -\omega_m \left(\frac{2.405}{R}\right)^2 \cdot k_{zi}^{-3}. \qquad (3.212)$$

An extensive investigation of the characteristic equation (3.206b) for both surface and volume magnetostatic modes can be found in the work of Joseph and Schlomann [45].

3.23.9   FARADAY ROTATION–CIRCULAR POLARIZATION (CP)

It is interesting to observe the azimuthal index (n) which is included in the third term of the characteristic equation (3.206b). The above index may be positive or negative. Recall that the azimuthal dependence $e^{jn\varphi}$ is responsible for circular polarization. A positive index $n = +1, +2, +3, \ldots$ corresponds to wave propagation toward negative $\hat{\varphi}$, which defines left-hand circular polarization (LHCP). Likewise, a negative index $n = -1, -2, -3, \ldots$ denotes wave propagation toward positive $\hat{\varphi}$ or right-hand circular polarization (RHCP). Instead, $n = 0$ modes represent linearly polarized waves. In turn, a different sign of index n (e.g., $n = +1, -1$) causes (3.206b) to yield a different axial wavenumber $\beta = k_{zi}$ as $\beta_+ \neq \beta_-$. This phenomenon can be interpreted as a "Faraday effect" for magnetostatic waves, which presents behavior similar to that of electromagnetic wave modes in ferrimagnetics. It can also be proved that propagation in the opposite direction (e.g., change of β_+, β_- signs) retains the same rotation angle; in other words, magnetostatic modes with $n \neq 0$ are non-reciprocal. This, in turn, enables a class of important applications of disks (e.g., in film form) and rods in non-reciprocal microwave devices, which are a unique feature of ferrimagnetics. Moreover, an opposite DC bias direction (e.g., changing the sign of an electromagnet's DC current) changes the sign of M_S and H_{DC} and results in a change of sign in the rotation angle.

To conclude this section, it is important to remember that circularly polarized waves, which enable the corresponding applications, are a direct consequence of the boundary conditions in the radial direction. Therefore, the radius of the sample determines their behavior.

3.24   MAGNETOSTATIC WAVES ON MULTILAYER AND GROUNDED STRUCTURES

The most attractive practical applications refer to miniaturized printed microwave devices [47, 48]. Working toward this trend, we will next examine useful structures of grounded dielectric-ferrite layers. Note that one of the most promising but still under investigation structures is ferrite-ferroelectric multilayer. The above will be qualitatively studied.

3.24.1   GROUNDED FERRITE SLAB

Seshadri [48] originally studied the surface magnetostatic modes guided by a longitudinally magnetized ferrite slab with one side metalized and grounded. The modes' characteristics were compared against those of the ungrounded slab given by Damon and Eshbach [13]. The geometry is similar to that of Figure 3.12; at the bottom side $x = 0$ is assumed metalized and grounded as shown in Figure 3.28.

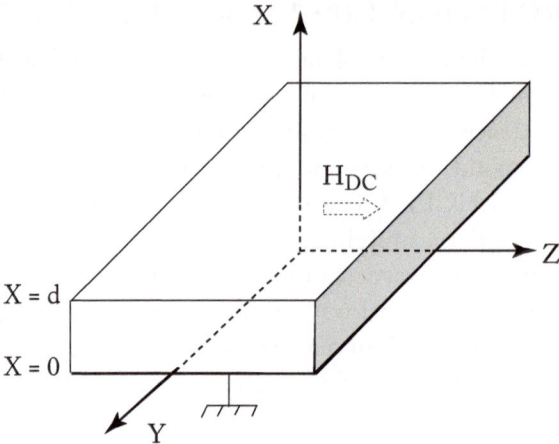

Figure 3.28: An infinitely extending longitudinally magnetized ferrite slab.

The general solution for the magnetic potential can be expressed with the aid of Eqs. (3.115) and (3.147) for volume and surface magnetostatic modes as follows:

$$\text{Ferrite region:} \quad \psi_i = f_i(x)e^{jk_y y}e^{-jk_z z} \quad 0 \le x \le d \tag{3.213a}$$

$$\text{Air region:} \quad \psi_e = \psi_{0e}e^{-\alpha(x-d)}e^{jk_y y}e^{-jk_z z} \quad x \ge d. \tag{3.213b}$$

where the appropriate eigenfunctions in the x-direction are:

$$\text{MS-volume modes:} \quad f_i(x) = f_{iv}(x) = A_x \sin(k_{xi}x) + B_x \cos(k_{xi}x) \tag{3.214a}$$

$$\text{MS-surface modes:} \quad f_i(x) = f_{is}(x) = A'_x e^{\alpha_i x} + B'_x e^{-\alpha_i x}. \tag{3.214b}$$

Also, recall that from the separation of variables Eq. (3.123) should be valid:

$$\text{Ferrite:} \quad (1 + \mathcal{X})k_{xi}^2 + (1 + \mathcal{X})k_y^2 + k_z^2 = 0 \tag{3.215a}$$

$$\text{Air:} \quad -\alpha^2 + k_y^2 + k_z^2 = 0. \tag{3.215b}$$

Also, the wavenumbers in the x-direction are denoted as:

$$\text{MS-volume modes:} \quad k_{xe}^2 = -\alpha^2 < 0 \quad \text{or} \quad k_{xe} = j\alpha \tag{3.216a}$$

$$\text{MS-surface modes:} \quad k_{xe}^2 = -\alpha^2 < 0 \quad \text{and} \quad k_{xi}^2 = -\alpha_i^2 \quad \leftrightarrow \quad k_{xi} = j\alpha_i. \tag{3.216b}$$

Propagation in either positive or negative y-direction is possible; that is, both signs of k_y are allowed.

The unknown constants in the general solutions of (3.214b) as well as the magnetostatic modes characteristic equations can be obtained through the application of the boundary conditions at $x = 0, d$. The familiar tangential magnetic field (h_y, h_z) and the normal magnetic flux density (b_x) continuity conditions should be enforced on the ferrite-air interface at $x = d$. The question is "what is the boundary condition for the magnetic field at the metallic (assumed perfect electric) ground plane at $x = 0$." This basic question is well established in classical electrodynamics that covers magnetostatics, e.g., Jackson [27] who states that:

"At the surface of a perfect electric conductor the magnetic field should obey the boundary conditions $\bar{B} \cdot \hat{n} = 0$ and $\hat{n} \times \bar{H} = \bar{J}_S =$ surface current density."

Thus, for the problem of Figure 3.28 the $b_x = 0$ should be imposed. In turn, the required field expressions can be written as:

$$\bar{h} = -\nabla\psi \ : \ h_y = -\frac{\partial\psi}{\partial y} = -jk_y \cdot \psi \quad \text{and} \quad h_z = -\frac{\partial\psi}{\partial z} = jk_z \cdot \psi. \tag{3.217}$$

$$\bar{b} = [\mu]\bar{h} = -[\mu]\nabla\psi \quad \leftrightarrow \quad b_{xi} = \mu_0(1 + \mathcal{X})h_{xi} + j\mu_0 k_1 \cdot h_{\psi i}. \tag{3.218a}$$

Then:

$$b_x = \begin{cases} -\mu_0\{(1 - \mathcal{X})f_i'(x) - k_1 k_y f_i(x)\}e^{jk_y y}e^{-jk_z z} & 0 \le x \le d \\ -\mu_0 \cdot (-\alpha) \cdot \psi_e = \mu_0 \cdot \alpha \cdot \psi_e & x \ge d. \end{cases} \tag{3.218b}$$

The conditions at the metallic ground plane $x = 0$ will be enforced separately for volume and surface modes, while the continuity conditions at the ferrite-air interface can be formulated for both types of modes. First continuity of either h_y or h_z asks for ψ continuity as:

$$f_i(x = d) = \psi_{0e} \tag{3.219a}$$

while b_x continuity yields:

$$(1 + \mathcal{X})f_i'(x = d) - k_1 k_y f_i(x = d) = -\alpha \cdot \psi_{0e}. \tag{3.219b}$$

The combination of the above gives:

$$(1 + \mathcal{X})f_i'(x = d) = (k_1 k_y - \alpha) \cdot f_i(x = d). \tag{3.219c}$$

This expression will be adapted to each particular mode type below.

3.24.2   MAGNETOSTATIC VOLUME MODES: $(1 + \mathcal{X}) < 0$

As for the case of the ferrite slab, for (3.215a) to be valid for real k_{xi}, k_y, and k_z, $(1 + \mathcal{X})$ should be negative. This requirement defines the range of magnetostatic volume modes given in (3.128b). Starting from the remaining boundary condition at the metallic ground plane, the equations (3.218b) and (3.213) yield:

$$b_x|_{x=0} = 0 \rightarrow (1 + \mathcal{X}) f_i'(x = 0) = k_1 k_y f_i(x = 0)$$

$$\leftrightarrow \quad (1 + \mathcal{X}) k_{xi} A_x = k_1 k_y B_x \tag{3.220a}$$

$$\text{or} \quad B_x = \frac{(1 + \mathcal{X})}{k_1} \cdot \frac{k_{xi}}{k_y} \cdot A_x. \tag{3.220b}$$

Substitution of (3.220a) and (3.220b) into (3.219c) through (3.213) results to the characteristic equation for magnetostatic volume modes:

$$(1 + \mathcal{X}) \cdot k_{xi} \alpha \cdot \cot(k_{xi} d) - (1 + \mathcal{X})^2 \cdot k_{xi}^2 + k_1 k_y \alpha - k_1^2 k_y^2 = 0. \tag{3.221}$$

This characteristic equation almost resembles (3.129) or (3.131) obtained for the ungrounded ferrite slab. Based on the image principle, one would expect the modes supported by the grounded slab to be identical to the anti-symmetric modes (with reference to b_x) of an ungrounded slab with double thickness $(2d)$. In other words, modes with $b_x(x = -d) = -b_x(x = d)$, by symmetry ensure that $b_x(x = 0) = 0$. In order to clarify this issue, we consider the particular case when the wave propagates purely in the $\hat{z}$-direction, that is, for $k_y = 0$.

Volume modes propagating only in the z-direction ($k_y = 0$).
    When $k_y = 0$, Eqs. (3.215) are reduced as in (3.135):

$$k_y = 0 \quad \leftrightarrow \quad \alpha = k_{xe} = k_z \quad \text{and} \quad k_{xi} = k_z / \sqrt{-(1 + \mathcal{X})}. \tag{3.222}$$

Substituting (3.222) into (3.221) the characteristic equation is reduced to:

$$\cot\left(k_z d / \sqrt{-(1 + \mathcal{X})}\right) = -\sqrt{-(1 + \mathcal{X})}. \tag{3.223}$$

Equation (3.223) would be identical to (3.140c) for an ungrounded ferrite slab with double thickness. Recall that (3.142b) represents the odd modes ($n = 1, 3, 5, \ldots$) which have a sinusoidal $\sin(k_{xi} x)$ or anti-symmetric distribution in the x-direction. This exactly justifies what it was expected above. Note that (3.223) was the only solution obtained for the grounded slab, while a solution corresponding to the even modes of the ungrounded slab is not an option. It is interesting to seek the dominant mode ($n = 1$) solution of (3.223) which equivalently defines its resonant frequency. More important is its group velocity to check whether its "backward wave" character is retained.

3.24.3   SURFACE WAVE MODES: $(1 + \mathcal{X}) > 0$

Similar to the ungrounded case, surface wave modes are expected in the frequency range where $(1 + \mathcal{X}) > 0$, also defined in Eq. (3.133c). For their characteristics let us impose the metallic boundary condition at $x = 0$, through (3.218b) and (3.214b):

$$b_x|_{x=0} = 0 \quad \leftrightarrow \quad (1 + \mathcal{X}) \alpha_i \left(A'_x - B'_x \right) = k_1 k_y \left(A'_x + B'_x \right)$$

$$B'_x = \frac{(1 + \mathcal{X}) \alpha_i - k_1 k_y}{(1 + \mathcal{X}) \alpha_i + k_1 k_y}. \tag{3.224}$$

Substituting (3.224) into (3.219c) through (3.214) yields the characteristic equation for the magnetostatic surface wave modes.

$$\frac{(1 + \mathcal{X}) \cdot \alpha_i - k_1 k_y}{(1 + \mathcal{X}) \cdot \alpha_i + k_1 k_y} \cdot e^{-2\alpha_i d} = \frac{(1 + \mathcal{X}) \cdot \alpha_i - k_1 k_y + \alpha}{(1 + \mathcal{X}) \cdot \alpha_i + k_1 k_y - \alpha}. \tag{3.225a}$$

Equation (3.225a) can be reformulated as:

$$e^{-2\alpha d} = \frac{(1 + \mathcal{X})^2 \cdot \alpha_i^2 - k_1^2 k_y^2 + (1 + \mathcal{X}) \cdot \alpha_i \alpha + k_1 k_y \cdot \alpha}{(1 + \mathcal{X})^2 \cdot \alpha_i^2 + k_1^2 k_y^2 - (1 + \mathcal{X}) \cdot \alpha_i \alpha + k_1 k_y \cdot \alpha}. \tag{3.225b}$$

This should be solved in conjunction with the relations (3.216b). However, it is more interesting to elaborate on the more practical case when propagation is purely transverse to the magnetization direction-$\hat{z}$, that is, $k_z = 0$.

3.24.4   SURFACE WAVES PROPAGATING ONLY IN THE $\hat{y}$–DIRECTION $(k_z = 0)$

Propagation only in the y-direction is a limiting but most practical case. Starting from equation (3.215), we get the reduced one as in (3.151):

$$\alpha^2 = \alpha_i^2 = k_y^2 \quad \text{or} \quad \alpha_i = \alpha = |k_y| \quad \text{and} \quad k_y = \pm \alpha. \tag{3.226}$$

Note that $\alpha_i, \alpha > 0$ while both signs of k_y are possible.

Using (3.226) into (3.225b) or dividing all terms by k_y^2, the characteristic equation reads:

$$e^{-2\alpha d} = \frac{(1 + \mathcal{X})^2 - k_1^2 + (1 + \mathcal{X}) + k_1 \cdot S}{(1 + \mathcal{X})^2 - k_1^2 - (1 + \mathcal{X}) + k_1 \cdot S} \tag{3.227a}$$

where $S = k_y/|k_y| = +1$ or -1.

The above equation can be easily recast into the form given by Seshadri [48] after using the notation $\mu_1 = \mu_0(1 + \mathcal{X})$, $\mu_2 = \mu_0 k_1$ and $\mu = \mu_1^2 - \mu_2^2$ as:

$$e^{-2\alpha d} = e^{-2|k_y|d} = -\frac{\mu_1 + \mu_2 \cdot S + \mu}{\mu_1 - \mu_2 \cdot S - \mu}. \tag{3.227b}$$

Substituting $\mathcal{X}$ and k_1 in terms of Ω and Ω_H from (3.71) the characteristic equation (3.227b) reads:

$$e^{-2\alpha d} = \frac{-2\Omega^2 + \Omega \cdot S + (\Omega_H + 1)(2\Omega_H + 1)}{\Omega_H + \Omega \cdot S + 1} \tag{3.227c}$$

or

$$\alpha = \alpha_i = |k_y| = \frac{1}{2d} \ln \left[\frac{\Omega_H + \Omega \cdot S + 1}{-2\Omega^2 + \Omega \cdot S + (\Omega_H + 1)(2\Omega_H + 1)} \right]_{\Omega \neq \Omega_H}. \tag{3.227d}$$

Observe first that the wavenumber is inversely proportional to twice the grounded slab thickness ($2d$), indicating that these modes correspond to magnetostatic surface waves of an ungrounded slab with the double thickness. Next, recall the assumed propagation $e^{+jk_y y}$, which is directed toward negative $\hat{y}$ when k_y is positive. So, the sign of $S = k_y/|k_y| = \pm 1$ in Eqs. (3.227) denotes that the propagation toward negative and positive $\hat{y}$-axis has different characteristics. Also, note that the expressions (3.227) are similar to that of Seshadri [48], but the argument of the logarithm is inverted and there is a difference in the sign. This is due to the different temporal assumption $e^{-j\omega t}$ of [48] accompanied with propagation toward positive y-direction $e^{+jk_y y}$ when $k_y > 0$. Nevertheless, the conclusions must be essentially the same.

Let us now identify the frequency range of the modes. Remember that we seek magnetostatic surface modes which exist only when $(1 + \mathcal{X}) > 0$, that is, in the frequency range $\Omega > \sqrt{\Omega_H(\Omega_H + 1)}$, according to (3.133c). Further, the argument of the logarithm in (3.227) should be positive to obtain a valid k_y. In addition, these modes have a resonance $|k_y| \to \infty$ when the argument of the logarithm tends to zero (either zero numerator or zero denominator). With the aid of some algebra, the above yields the propagation characteristics identical to these of [48] as:

Toward positive: $\hat{y}$, $k_y < 0$, $S = -1$:

range $\Omega_1 = \sqrt{\Omega_H(\Omega_H + 1)} < \Omega_2 = \Omega_H + 0.5$ resonance at $\Omega = \Omega_2$. (3.228a)

Toward negative: $\hat{y}$, $k_y > 0$, $S = +1$:

range $\Omega_1 < \Omega < \Omega_3 = \Omega_H + 1$ resonance at $\Omega = \Omega_3$. (3.228b)

Seshadri [48] pointed out an important observation, which is obvious in (3.228a) and (3.228b); that is, that the range of MSSW for a longitudinally magnetized grounded ferrite slab extends to $\Omega_3 = \Omega_H + 1$, while for the ungrounded slab it was restricted to $\Omega_2 = \Omega_H + 0.5$, e.g., Damon and Eshbach [13], or (3.154).

The MSSW group velocity can be found by differentiating (3.227) with respect to Ω and inverting the result as follows:

$$V_g = \frac{d\omega}{dk_y} = \omega_m \frac{d(\omega/\omega_m)}{dk_y} = \omega_m \left(\frac{dk_y}{d\Omega} \right)^{-1}. \tag{3.229a}$$

After some algebra, this concludes to:

$$V_g = 2d \cdot \omega_m(-S) \cdot \frac{\left[\Omega^2 - (\Omega_H + 1)^2\right] \cdot [\Omega \cdot S + (\Omega_H + 0.5)]}{[\Omega^2 + 2\Omega \cdot S \cdot (\Omega_H + 1) + \Omega_H \cdot (\Omega_H + 1)]}. \tag{3.229b}$$

Note, that for this section to be compatible with the previous one and, in turn, with the work of Damon and Eshbach [13], the $\hat{y}$ propagation was assumed $e^{+jk_y y} = e^{j|k_y|S \cdot y}$. Thus, $S = -1$ represents propagation toward positive $\hat{y}$ for which (3.229b) yields positive V_g. That is, MSSW are ordinary forward modes with group velocity pointing toward the propagation direction. Also, it is worth noting that when the metallic ground plane is located on the upper face $x = d$ of the slab instead of the lower face $x = 0$, then the characteristics of MSSW propagating in the positive $\hat{y}$ and negative $\hat{y}$ directions are interchanged. An important result shown in (3.229) is that the group velocity V_g is proportional to the substrate thickness d. Since "delay lines" represent a very attractive application the desired group delay $t_d = \ell/V_g$ (ℓ is the structure length) is inversely proportional to the slab thickness $t_d \propto 1/d$. In other words, the slab thickness may be exploited as a degree of freedom in controlling the group delay.

Further, it is worth comparing the MSSW wavenumber and group velocity with those of the ungrounded ferrite slab, which from Eqs. (3.157b) and (3.158) [48], reads:

$$|k_y| = -\frac{1}{2d} \ln\left[4\left(\Omega_H + 0.5\right)^2 - \Omega^2\right]. \tag{3.230}$$

$$V_g = \frac{\omega_m d}{\Omega}\left[(\Omega_H + 0.5)^2 - \Omega^2\right]. \tag{3.231}$$

Comparing (3.231) with (3.229), one may see that in both cases the group delay ($t_d \propto 1/V_g$) is inversely proportional to the ferrite slab thickness. However, the sign (S) is absent from the ungrounded ferrite slab relations (3.230)–(3.231), meaning that MSSW propagating in both positive and negative y-directions have the same characteristics.

3.24.5   GROUNDED DIELECTRIC–FERRITE LAYERS

Continuing with the printed ferrite structures, the next to be studied is the grounded dielectric-ferrite layers shown in Figure 3.29 [47–50]. Bongianni [47], who in 1972 followed the notation of Seshadri [48], originally studied this. However, herein the notation and coordinate system of the previous sections will be used. The almost ideal dielectric substrate for YIG film growth is the gadolinium gallium garnet (GGG), as pointed out by Ishak [50]. The practical ferrite YIG film thickness ranges from 5–150 μm, with linewidth as low as $\Delta H \approx 0.5$ Oe.

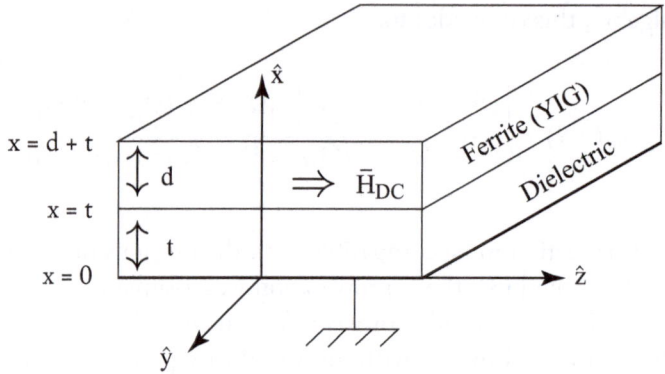

Figure 3.29: Infinite longitudinally magnetized two dielectric-ferrite layers.

Following the reasoning of the grounded ferrite section, the general solution for the magnetic potential can be written as:

$$\text{Dielectric region:}\quad \psi_d = f_d(x)e^{jk_y y}e^{-jk_z z} \tag{3.232a}$$

$$\text{Ferrite region:}\quad \psi_i = f_i(x)e^{jk_y y}e^{-jk_z z} \tag{3.232b}$$

$$\text{Air region:}\quad \psi_e = \psi_{0e}e^{-\alpha(x-d-t)}e^{jk_y y}e^{-jk_z z}. \tag{3.232c}$$

The appropriate eigenfunction in the x-direction is:

$$f_d(x) = A_d e^{k_{xd}x} + B_d e^{-k_{xd}x} \tag{3.233a}$$

$$\text{MS volume modes:}\quad f_i(x) = f_{ix}(x) = A_x \sin(k_{xi}x) + B_x \cos(k_{xi}x) \tag{3.233b}$$

$$\text{MS surface modes:}\quad f_i(x) = f_{is}(x) = A'_x e^{\alpha_i x} + B'_x e^{-\alpha_i x}. \tag{3.233c}$$

For the wavenumber relations, Equations (3.215) are still valid for the air and ferrite layers as:

$$\text{Air:}\quad -\alpha^2 + k_y^2 + k_z^2 = 0 \tag{3.234a}$$

$$\text{Ferrite:}\quad (1+\mathcal{X})k_{xi}^2 + (1+\mathcal{X})k_y^2 + k_z^2 = 0, \tag{3.234b}$$

while for the dielectric layer it is:

$$\text{Dielectric:}\quad -k_{xd}^2 + k_y^2 + k_z^2 = 0 \quad \leftrightarrow \quad k_{xd} = \alpha. \tag{3.234c}$$

A point of interest here is the expected field behavior in the dielectric region, which is directly assumed to be similar to that of surface waves, that is, an exponential decay away from the surface. This is dictated by the wavenumber relation (3.234) where k_y and k_z should be real, $(k_y^2, k_z^2 > 0)$, representing propagation along the infinite $\hat{y}$ and $\hat{z}$ directions. This asks k_x^2 to be negative for (3.234c) to be fulfilled as $k_x^2 < 0$ or $k_x^2 = -k_{xd}^2$ and $k_x = jk_{xd}$. Comparing (3.234a) and (3.234c), it is obvious that $k_{xd} = \alpha$. That is, the wave behavior in the air and dielectric are quite similar. This is intuitively expected since the dielectric is a non-magnetic material $(\mu_d = \mu_0)$. It is also expected that when the dielectric thickness tends to infinity $(t \to \infty)$, then the wave characteristics will tend to those of the ungrounded (free) ferrite slab. Likewise, when its thickness tends to zero $(t \to 0)$, the structure will behave exactly as the grounded ferrite slab.

Let us now seek the magnetostatic volume or surface-wave modes characteristic equations through the enforcement of the boundary conditions. Starting from the ferrite-air interface $x = d + t$, the continuity conditions remain the same as in the previous section. Since the same notation is retained, we may directly adapt (3.219c) to this case as:

$$(1 + \mathcal{X}) f_i'(x = d + t) = (k_1 k_y - \alpha) \, f_i(x = d + t). \tag{3.235}$$

The normal magnetic flux density should again vanish at the metallic ground plane at $x = 0$:

$$b_{xd} = \mu_0 h_{xd} = -\mu_0 \frac{\partial \psi_d}{\partial x} \bigg|_{x=0} = 0 \quad \leftrightarrow \quad f_d'(x = 0) = 0 \quad \leftrightarrow \quad B_d = A_d. \tag{3.236}$$

Substituting (3.236) to (3.233a), we have:

$$f_d(x) = 2A_d \cosh(k_{xd} x) \quad \text{and} \quad f_d'(x) = 2A_d \cdot k_{xd} \cdot \sinh(k_{xd} x). \tag{3.237}$$

The normal flux density in the ferrite region is given in (3.218b), so the continuity condition for h_y, h_z, and b_x yields correspondingly:

$$f_d(x = t) = f_i(x = t) \tag{3.238a}$$

$$(1 + \mathcal{X}) \, f_i'(x = t) = -k_1 k_y f_i(x = t) = -f_d'(x = t). \tag{3.238b}$$

Combine the above together through (3.237) to get:

$$(1 + \mathcal{X}) f_i'(x = t) = \left[k_1 k_y - k_{xd} \tanh(k_{xd} t) \right] f_i(x = t). \tag{3.239}$$

Equations (3.235) and (3.239) can be readily combined in conjunction with the eigenfunctions of (3.233) to yield the characteristic equations of either volume or surface magnetostatic modes.

3.24.6   MAGNETOSTATIC VOLUME MODES $(1 + \mathcal{X}) < 0$

The eigenfunction $f_i(x)$ from (3.233b) is substituted in (3.235), (3.239), and the A_x, B_x terms are separated. In turn, dividing the resulting expressions, we end up with the dispersion equation:

$$\left. \frac{(1 + \mathcal{X})k_{xi} - (k_1 k_y - \alpha) \tan [k_{xi}(d + t)]}{(1 + \mathcal{X})k_{xi} - [k_1 k_y - k_{xd} \tan (k_{xd}t) \tan (k_{xi}t)]} \right|_{k_{xd}=\alpha}$$

$$= \left. \frac{(1 + \mathcal{X})k_{xi} \tan [k_{xi}(d + t) + (k_1 k_y - \alpha)]}{(1 + \mathcal{X})k_{xi} \tan (k_{xi}t) + [k_1 k_y - k_{xd} \tan (k_{xd}t)]} \right|_{k_{xd}=\alpha} . \qquad (3.240\text{a})$$

The above dispersion equation can be simplified again by performing the multiplications and exploiting the tangent addition identity to yield:

$$\tan (k_{xi}d) = \frac{(1 + \mathcal{X})k_{xi} \cdot \alpha \cdot [1 - \tanh(\alpha t)]}{(1 + \mathcal{X})^2 k_{xi}^2 + k_1^2 k_y^2 - \alpha \cdot k_1 k_y [1 + \tanh(\alpha t)] + \alpha \cdot \tanh(\alpha t)} . \qquad (3.240\text{b})$$

In the limiting case of infinite dielectric thickness $(t \to \infty)$, the hyperbolic tangent $\tanh(k_{xd}t) \to 1$ and (3.240) reduces to (3.129) obtained for the ferrite slab. On the other hand, when the dielectric thickness tends to zero $t \to 0 \leftrightarrow \tanh(k_{xd}t) \approx (k_{xd}t) = \alpha t \to 0$ and (3.240) becomes identical to (3.220b), the one for the grounded ferrite slab. As before, let us examine the case of pure propagation in the z-direction.

3.24.7   VOLUME MODES PROPAGATING ONLY IN THE z–DIRECTION $(k_y = 0)$

For the case where $k_y = 0$ and $\bar{k}_i = k_z \hat{z}$, the wavenumbers are related through (3.222). Combining it with (3.240) and exploiting the tangent addition formula, e.g., Abramowitz [39], after some algebra we end up with the following dispersion relation:

$$\tan \left[\frac{k_z d}{\sqrt{-(1 + \mathcal{X})}} \right] = \frac{\sqrt{-(1 + \mathcal{X})}}{(1 + \mathcal{X}) - \tanh(k_z t)} [1 - \tanh (k_z t)] . \qquad (3.241)$$

It is worth remembering that these are backward modes (MSBVW) since they are obtained from a longitudinal bias. In other words, their group velocity $(V_g = d\omega/dk_z)$ can be proved to be negative.

3.24.8   MAGNETOSTATIC SURFACE MODES $(1 + \mathcal{X}) > 0$

In the case of $(1 + \mathcal{X}) > 0$, the eigenfunction $f_i(x)$ from (3.233c) is substituted to (3.235) and (3.239). Likewise, the terms involving A'_x and B'_x are separated, and the resulting expressions are divided to yield the dispersion equation.

$$\frac{\left[(1+\mathcal{X})\alpha_i - (k_1 k_y - \alpha)\right]e^{\alpha_i d}}{(1+\mathcal{X})\alpha_i - \left[k_1 k_y - k_{xd}\tan{(k_{xd}t)}\right]}\Bigg|_{k_{xd}=\alpha}$$

$$= \frac{\left[(1+\mathcal{X})\alpha_i + (k_1 k_y - \alpha)\right]e^{-\alpha_i d}}{(1+\mathcal{X})\alpha_i + \left[k_1 k_y - k_{xd}\tan{(k_{xd}t)}\right]}\Bigg|_{k_{xd}=\alpha}. \tag{3.242}$$

As expected, in the limiting case of $(t \to \infty)$, Eq. (3.242) is reduced to (3.150) for the ungrounded slab, while for $(t \to 0)$, the dispersion equation yields to the grounded ferrite slab (3.225a). As in the previous sections, we will focus on the practical case of pure propagation along the $\hat{y}$-direction.

3.24.9 SURFACE WAVES PROPAGATING ONLY IN THE $\hat{y}$–DIRECTION $(k_z = 0)$

In the particular case where $(k_z = 0)$, the wavenumbers obey Eq. (3.226), namely:

$$k_{xd} = \alpha = \alpha_i = |k_y| \text{ and } k_y = \pm\alpha \text{ also } S = \frac{k_y}{|k_y|} = \pm 1 \text{ or } k_y = |k_y| \cdot S. \tag{3.243}$$

In view of the above, (3.242) reads:

$$e^{-2a_i d} = \frac{[(1+\mathcal{X}) - k_1 S + 1]\left[(1+\mathcal{X}) + k_1 S - \tanh{\left(|k_y|t\right)}\right]}{[(1+\mathcal{X}) + k_1 S + 1]\left[(1+\mathcal{X}) - k_1 S + \tanh{\left(|k_y|t\right)}\right]}. \tag{3.244}$$

The above expression is in agreement with that of O'Keefe and Patterson [49], just by using the corresponding notation. In the limiting case of zero dielectric thickness $(t \to 0)$ equation (3.244) is reduced to the grounded ferrite case (3.227a) just performing the multiplications. Likewise, when $(t \to \infty)$, Eq. (3.244) becomes identical to (3.150) obtained for the ungrounded ferrite case. To show the agreement with Bongianni's expression [47], one may solve (3.242) for the hyperbolic tangent as:

$$\tanh{\left(|k_y|t\right)} = (1+\mathcal{X})\left\{e^{2|k_y|d} - \frac{(1+\mathcal{X}) + k_1 S - 1}{(1+\mathcal{X}) - k_1 S + 1}\right\}$$

$$\left\{e^{2|k_y|d} + \frac{(1+\mathcal{X}) + k_1 S - 1}{(1+\mathcal{X}) - k_1 S + 1}\right\} + k_1 S. \tag{3.245}$$

Expression (3.245) becomes identical to Bongianni's [47], by letting $\mu_1 = \mu_0(1 + \mathcal{X})$, $\mu_2 = \mu_0 k_1$, and $S \to -S$, as explained in the previous section for equation (3.227b). Notice, the presence of sign $S = \pm 1$ in the dispersion equation causes the modes propagating toward positive and negative $\hat{y}$-direction to have different characteristics, just like the grounded ferrite slab.

3.24.10 MAGNETOSTATIC MODES OF A FINITE WIDTH SLAB

A point of practical interest is the investigation of magnetostatic modes characteristics when the ferrite slab (film) has a finite width (w). These modes have been studied by O'Keefe and Patterson [49], and presented also by Ishak [50]. Herein, it can be relatively easy included in the above analysis. Simply, a slight modification of the eigenfunctions given in Eqs. (3.232) is needed. First, we must decide which of the two infinite slab dimensions in $\hat{y}$ or $\hat{z}$-directions will be restricted to a finite width w. This is directly related to the type of magnetostatic waves to be sought, since for a $\hat{z}$-biased slab, volume modes are primarily propagating in the $\hat{z}$-direction, while surface modes are primarily directed along $\hat{y}$-direction. Let us study both in turn.

3.24.11 SURFACE MODES OF FINITE WIDTH SLAB $(-w/2 \leq z \leq w/2)$

Propagation along an infinite $\hat{y}$-direction as $e^{jk_y y}$ will be retained. In addition, a standing wave should be considered along the finite z-dimension. For this purpose, let us replace the corresponding eigenfunctions in (3.232) as:

$$e^{-jk_z z} \; \rightarrow \; h(z) = A_z \cdot e^{-jk_z z} + A_z \cdot e^{+jk_z z}. \tag{3.246}$$

All the previous boundary conditions applied on $f(x)$ eigenfunctions should be preserved as they are, since $h(z)$ appears the same in all different x-ranges. However, new boundary conditions should be imposed at the two edges $z = \pm w/2$, that is, at the new ferrite (or material)—air interfaces. Unfortunately, this boundary condition is not obvious and there is not any convenient exact expression. The good news is that *a very simple approximate "pinning" condition was considered and tested against measured samples*. In [48], it is stated that: "The sample edges at $z = \pm w/2$ are practically defined by slicing or chemical etching." The combination of the edge roughness produced by the sample preparation and the demagnetizing fields are enough to "*pin the spins at the sample edges.*" In turn, the pinning condition is approximated therein by assuming the tangential components of the flux density (b_x, b_y) vanish at the edges $z = \pm w/2$. These components are defined by using (3.72) and (3.73):

$$b_x = \mu_0 \left[(1 + \mathcal{X}) h_x + j k_1 h_y \right] = -\mu_0 \left[(1 + \mathcal{X}) \partial \psi / \partial x + j k_1 \partial \psi / \partial y \right] \tag{3.247a}$$

$$b_y = \mu_0 \left[-j k_1 h_x + (1 + \mathcal{X}) h_y \right] = -\mu_0 \left[-j k_1 \partial \psi / \partial x + (1 + \mathcal{X}) \partial \psi / \partial y \right]. \tag{3.247b}$$

In view of (3.232) through (3.246), both (3.247) should be forced to vanish at the edges by requiring $h(z) = $ zero:

$$h(z = \pm w/2) = 0 \; \rightarrow \; k_z w = n\pi \text{ or } k_z = n\pi/w \text{ where } n = 0, 1, 2, \tag{3.248}$$

and

$$h(z) = -2j A_z e^{-jk_z w/2} \cdot \sin \left[k_z (z - w/2) \right] = A_z' \sin \left[\frac{n\pi}{w} \left(z - \frac{w}{2} \right) \right]. \tag{3.249}$$

Note that the term $e^{-jk_z w/2} = e^{-jn\pi/2}$ is constant and is thus absorbed in the constant A'_z. The wavenumbers are still given by (3.234), which in view of (3.248) and adopting the notation of [49] for direct comparison purposes, becomes:

$$N^2 = k_{xd}^2 = \alpha^2 = k_y^2 + k_z^2 = k_y^2 + (n\pi/w)^2 \tag{3.250a}$$

$$M^2 = -k_{xi}^2 = a_i^2 = k_y^2 + \frac{k_z^2}{(1 + \mathcal{X})} = k_y^2 + \frac{(n\pi/w)^2}{(1 + \mathcal{X})}. \tag{3.250b}$$

Remember that we are dealing with surface modes occurring when $(1 + \mathcal{X}) > 0$. Their dispersion equation (3.242) through (3.250) reads:

$$e^{-2Md} = \frac{(1 + \mathcal{X})M - k_1 \cdot k_y + N}{(1 + \mathcal{X})M + k_1 \cdot k_y - N} \cdot \frac{(1 + \mathcal{X})M + k_1 \cdot k_y - N \tan(Nt)}{(1 + \mathcal{X})M - k_1 \cdot k_y + N \tan(Nt)}. \tag{3.251}$$

In [49], the notation $(1 + \mathcal{X}) = \mu_1$, $k_1 = \mu_2$ and $k_y = k$ is used, while therein the structure is modeled upside down accessing infinity at $x = -\infty$. The latter justifies the different signs observed in (3.251). Also [49], a question that is raised is about the slab-width effects. When are they significant and when could they be ignored? A qualitative answer could say that width effects are significant when the wavelength $\lambda_y = 2\pi/k_y$ is comparable or greater to the sample width ($\lambda_y = 2\pi/k_y \geq w$). On the contrary, in the limit when $\lambda_y << w$, a behavior like the infinite film (3.242) is expected. A corresponding practical limit is set in [49]. Hence, the limiting wavenumber is set to be $k_w = 10$ n/w. When $k_y > k_w$ or $\lambda_y < 0.628$ w/n, width effects are ignored and vice versa. Further, according to [49], volume modes may occur in the range of $(1 + \mathcal{X}) < 0$.

3.24.12 VOLUME MODES OF A SLAB FINITE IN THE z–DIMENSION

As explained in the previous section, volume modes tend to propagate along the direction of dc-magnetization ($\hat{z}$-axis) with $|k_y| \to 0$ and $|k_i| \approx k_z$. However, for the finite slab truncated in the z-direction, there could be a small frequency range within $(1 + \mathcal{X}) < 0$ where volume modes occur. Remember that $(1 + \mathcal{X}) < 0$ allows for all three wavenumbers k_{xi}, k_y, and k_z to be real (in the lossless case). In turn, expressions (3.248) through (3.250a) are valid, while (3.250b) is modified as:

$$k_{xi}^2 = M^2 = \frac{(n\pi/w)^2}{-(1 + \mathcal{X})} - k_y^2 > 0. \tag{3.252}$$

The dispersion equation is given by (3.222), which is specialized for the present case by substituting for k_{xi} and $k_{xd} = \alpha$ from (3.252) and (3.250a), respectively.

Another point of interest stated in [49] is the possibility of surface waves (MSSW) to become backward, i.e., to have opposite phase and group velocities.

3.24.13 VOLUME MODES OF A SLAB FINITE IN y–DIMENSION

A logical question could be raised as to why not study volume modes propagating in the z-direction as $e^{-jk_z z}$ of slab finite in its y-dimension, and thus with a standing-wave pattern along the y-axis. This is actually a very interesting case which is already exploited in practice, but for the electromagnetic modes, also known as "edge modes." These modes have mode patterns cling to either $y = -w/2$ or $y = +w/2$ and propagating in opposite z-directions. So, they are expected to behave like the magnetostatic surface modes which cling to either top $x = d/2$ or bottom $x = -d/2$ face of the ungrounded slab. Learning from the latter situation, one intuitively expects that a non-zero magnetic potential beyond $y = \pm w/2$ should be considered. However, this becomes far too complicated to be analyzed herein. In order to realize the complication, let us try to apply the "pinning" conditions of the previous section. That is, assume sample edges at $y = \pm w/2$ and try to enforce the tangential flux density components b_x and b_z to be zero on them. Consider a standing-wave eigenfunctions along y-axis as:

$$g(y) = A_y e^{-jk_y y} + B_y e^{+jk_y y}. \tag{3.253a}$$

Pinning b_z at $y = \pm w$ requires:

$$b_z = \mu_0 h_z = -\mu_0 \partial \psi / \partial z = \mu_0 (-jk_z) \psi |_{y=\pm w/2} = 0.$$

Then

$$g(y = \pm w/2) = 0 \rightarrow k_y \cdot w = m\pi; \qquad m = 0, 1, 2, \ldots \tag{3.253b}$$

$$\text{and} \quad g(y) = A'_y \sin \left[\frac{m\pi}{w} \left(y - \frac{w}{2} \right) \right]. \tag{3.253c}$$

However, b_x is also tangential and must be pinned at the edges. Thus:

$$b_x = -\mu_0 \left[(1 + \mathcal{X}) \frac{\partial \psi}{\partial x} + jk_1 \frac{\partial \psi}{\partial x} \right]_{y=\pm w/2} = 0. \tag{3.254a}$$

This in turn asks for:

$$\left[(1 + \mathcal{X}) f'_i(x) g(y) + jk_1 f_i(x) g'(y) \right]_{y=\pm w/2} = 0. \tag{3.254b}$$

In view of (3.253c), the above requires:

$$g'(y = \pm w/2) = \frac{m\pi}{w} A'_y \cos \left[\frac{m\pi}{w} \left(y - \frac{w}{2} \right) \right]_{y=\pm w/2} = 0. \tag{3.254c}$$

The above has only one solution, which is $m = 0$, or the potential ψ to be zero everywhere. Hence, *the pinning conditions are unacceptable in this case.*

Let us now study the magnetostatic forward volume of the finite grounded structure of Figure 3.30.

3.25 TRANSVERSELY BIASED GROUNDED DIELECTRIC–FERRITE LAYERS

The aim of this section is the analysis of the structure of Figure 3.30 with a normal DC biasing field. As a matter of convenience, the coordinate system is oriented as shown in Figure 3.30.

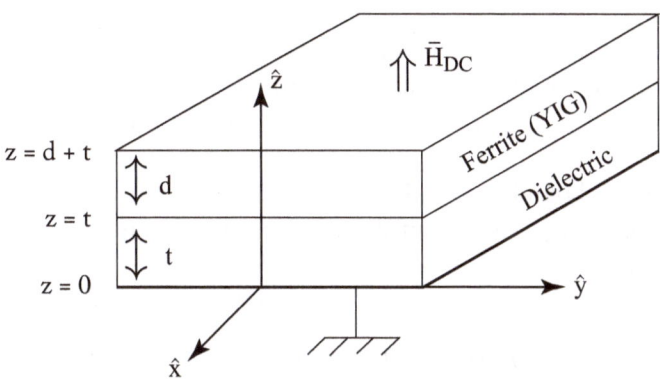

Figure 3.30: Transversely magnetized dielectric-ferrite double layer (demagnetization $H_i = H_{DC} - M_S$).

It would be interesting to study directly either a finite sample in $\hat{x}$ or in $\hat{y}$-directions. However, this involves the same difficulties discussed for volume modes in the previous sections. Thus, the layers will be considered infinite in both $\hat{x}$ and $\hat{y}$-directions for convenience, assuming propagation as $e^{-j\bar{k}_t \cdot \bar{\rho}}$ as in Eqs. (3.89). In turn, the magnetic potential general solutions in the three regions take the form:

$$\text{Dielectric region:} \quad \psi_d = h_d(z)e^{-j\bar{k}_t \cdot \bar{\rho}} \quad \text{for } 0 \leq z \leq t \tag{3.255a}$$

$$\text{Ferrite region:} \quad \psi_i = h_i(z)e^{-j\bar{k}_t \cdot \bar{\rho}} \quad \text{for } t \leq z \leq d + t \tag{3.255b}$$

$$\text{Air region:} \quad \psi_e = h_{0e}e^{-j\bar{k}_t \cdot \bar{\rho}} \, e^{-\alpha(z-d-t)} \quad \text{for } z \geq d + t, \tag{3.255c}$$

where the z-eigenfunctions are:

$$h_d(z) = A_d e^{k_{zd} \cdot z} + B_d e^{-k_{zd} \cdot z} \tag{3.256a}$$

$$h_i(z) = A_z \sin(k_{zi}z) + B_z \cos(k_{zi}z), \tag{3.256b}$$

with wavenumbers satisfying the relations (3.90) as:

$$\text{Dielectric:} \quad k_t^2 - k_{zd}^2 = 0 \leftrightarrow k_{zd}^2 = k_t^2 \tag{3.257a}$$

$$\text{Ferrite:} \quad (1 + \mathcal{X})k_t^2 + k_{zi}^2 = 0 \leftrightarrow k_{zi}^2 = -(1 + \mathcal{X})k_t^2 \tag{3.257b}$$

$$\text{Air:} \quad k_t^2 - \alpha^2 = 0 \leftrightarrow \alpha^2 = k_t^2 = k_{zd}^2. \tag{3.257c}$$

Note that for the wavenumber relations to be satisfied, an exponentially attenuated eigenfunction $h(z)$ is assumed in both the air and dielectric regions. On the contrary, the possibility of the relative permeability $(1 + \mathcal{X})$ to become negative allows wave propagation (or $k_{zi}^2 > 0$) in the ferrite region.

Let us enforce the boundary conditions to find the dispersion equation. First, the normal magnetic flux density (b_z) should vanish at the metallic ground plane. Similarly to (3.236) and (3.237), one gets:

$$h_d(z) = 2A_d \cosh(k_{zd}z) \tag{3.258a}$$

$$h_d'(z) = 2A_d \cdot k_{zd} \cdot \sinh(k_{zd}z). \tag{3.258b}$$

Next, the continuity conditions of the normal flux density (b_z) and tangential components of the magnetic field h_x and h_y should be imposed at the interfaces: dielectric-ferrite at $z = t$ and ferrite-air at $z = d + t$. Upon inspecting Equations (3.91) and (3.92), it suffices to enforce the continuity of $\partial \psi / \partial z$ and $\psi(z)$, respectively. After some algebra to remove A_d and ψ_{0e}, these conditions yield:

$$z = t : k_{zd} \tanh(k_{zd}t) [A_z \tanh(k_{zi}t) + B_z] = k_{zi} [A_z - B_z \tanh(k_{zi}t)] \tag{3.259a}$$

$$z = d + t : k_{zi} [A_z - B_z \tan[k_{zi}(d + t)]] = -\alpha [A_z \tan[k_{zi}(d + t)] + B_z]. \tag{3.259b}$$

The dispersion equation can be readily obtained by separating A_z, B_z terms and dividing the two resulting equations, as:

$$\frac{k_{zi} - k_{zd} \tanh(k_{zd}t) \cdot \tan(k_{zi}t)}{k_{zi} \cdot \tan(k_{zi}t) + k_{zd} \cdot \tanh(k_{zd}t)} = \frac{k_{zi} + \alpha \cdot \tan[k_{zi}(d + t)]}{k_{zi} \cdot \tan[k_{zi}(d + t)] - \alpha} \tag{3.260a}$$

where $k_{zd} = \alpha = k_t$ and $k_{zi} = k_t \sqrt{-(1 + \mathcal{X})}$.

Equation (3.260a) is simplified by performing the multiplications and using the tangent addition formula to give:

$$\tan\left(k_t d \sqrt{-(1 + \mathcal{X})}\right) = -\frac{\sqrt{-(1 + \mathcal{X})} \cdot [1 + \tanh(\alpha t)]}{(1 + \mathcal{X}) + \tanh(\alpha t)}. \tag{3.260b}$$

In the limiting case when the dielectric is absent $(t \to 0)$, then (3.260b) should give the dispersion equation of a transversely biased grounded ferrite slab as:

$$\text{Grounded ferrite slab:} \quad k_{zi} \tan(k_{zi}d) = \alpha \tag{3.261a}$$

$$\text{or} \quad \tan(k_t d \sqrt{-(1+\mathcal{X})}) = \frac{1}{\sqrt{-(1+\mathcal{X})}}. \tag{3.261b}$$

These are volume modes occurring in the frequency range $(1 + \mathcal{X}) < 0$, or according to (3.131) for $\Omega_H < \Omega < \sqrt{\Omega_H(\Omega_H + 1)}$. The above dispersion equation is identical to that for "symmetric modes" of an ungrounded ferrite slab with double thickness $(2d)$, as given in (3.98) or (3.102). Note that the term "symmetric" refers to the cosine dependence of the magnetic potential $\psi(-z) = \psi(z)$, which corresponds to "anti-symmetric" or sinusoidal dependence for the normal flux density $b_z = -\mu_0 \partial \psi / \partial z$ or $b_z(-z) = b_z(z)$. This result is also justified by the "image principle." An also interesting structure is the shielded one, which constitutes the next task.

3.26   SHIELDED DIELECTRIC–FERRITE LAYERS

In most practical applications, the microwave devices are enclosed within a metallic box. The structure is shown in Figure 3.31. In turn, this will be studied for both transverse and longitudinal DC magnetization. Pareck et al. [51] also present these shielded structures.

3.26.1   TRANSVERSELY MAGNETIZED SHIELDED DIELECTRIC–FERRITE LAYERS (MSFVW)

For the analysis of the structure of Figure 3.31, the formulation of the previous section can be exploited. The only additional difference concerns the z-eigenfunction in the air region, where a wave "reflected" from the metallic screen located at $z = z_2$ will be added. For this purpose, the eigenfunction ψ_e in the air region is modified for standing-wave behavior as:

$$\psi_e = h_e(z)e^{-j\bar{k}_t \cdot \bar{\rho}} \quad \text{where} \quad h_e(z) = A_e e^{-\alpha z} + B_e e^{\alpha z}. \tag{3.262}$$

The boundary condition at the metallic shield $z = z_2$ asks for the vanishing of the normal component of the magnetic flux density:

$$b_z = -\mu_0 \frac{\partial \psi}{\partial z} \quad \leftrightarrow \quad \frac{dh_e(z = z_2)}{dz} = h'_e(z = z_2) = 0 \quad \leftrightarrow \quad B_e = A_e e^{-2\alpha z_2}. \tag{3.263a}$$

So, the $h_e(z)$ eigenfunctions should be modified as:

$$h_e(z) = 2A_e e^{-\alpha z_2} \cosh[\alpha(z - z_2)] = A'_e \cosh[\alpha(z - z_2)] \tag{3.263b}$$

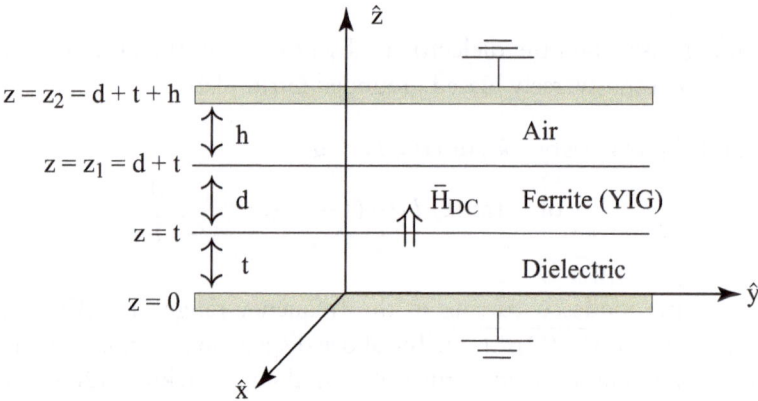

Figure 3.31: Shielded transversely magnetized dielectric-ferrite double layers.

and

$$h'_e(z) = \alpha \cdot A'_e \cdot \sinh\left[\alpha \left(z - z_2\right)\right]. \tag{3.263c}$$

The boundary conditions at $z = 0$ and $z = t$ are preserved as in the previous section and end up to (3.259a) to be used herein. Now, the continuity of ψ and $\partial\psi/\partial z$ should be enforced at the ferrite air interface at $z = z_1 = d + t$. These yield the relation:

$$\frac{A_z \tanh\left(k_{zi}z\right) + B_z}{k_{zi}\left[A_z - B_z \tanh\left(k_{zi}z\right)\right]} = \frac{1}{\alpha \cdot \tanh\left[\alpha\left(z_1 - z_2\right)\right]} = \frac{1}{\alpha \cdot \tanh\left[\alpha h\right]}. \tag{3.264}$$

The combination of (3.259a) with (3.264), including the separation of A_z, B_z terms and the division of the resulting expressions, yields the dispersion equation:

$$\frac{k_{zi} - k_{zd}\tan\left(k_{zd}t\right)\tan\left(k_{zi}t\right)}{k_{zi}\tan\left(k_{zi}t\right) + k_{zd}\tan\left(k_{zd}t\right)} = \frac{k_{zi} + \alpha \cdot \tanh(\alpha h) \cdot \tan\left[k_{zi}(d + t)\right]}{k_{zi}\tan\left[k_{zi}(d + t)\right] - \alpha \cdot \tanh(\alpha h)} \tag{3.265a}$$

where

$$k_{zd} = \alpha = k_t \quad \text{and} \quad k_{zi} = k_t\sqrt{-(1 + \mathcal{X})}. \tag{3.265b}$$

Equation (3.265b) can be greatly simplified by performing the multiplications and using the tangent addition formula to give:

$$\tan\left(k_{zi}d\right) = \frac{\alpha \cdot k_{zi}\left[\tanh(\alpha h) + \tanh(\alpha t)\right]}{k_{zi}^2 - \alpha^2 \cdot \tanh(\alpha h) \cdot \tanh(\alpha t)}. \tag{3.265c}$$

Substituting the wavenumber relations from (3.265b) and (3.257), the dispersion equation is further simplified:

$$\tan\left(k_t d\sqrt{-(1 + \mathcal{X})}\right)\Big|_{k_t = \alpha} = -\frac{\sqrt{-(1 + \mathcal{X})} \cdot \left[\tanh(\alpha h) + \tanh(\alpha t)\right]}{(1 + \mathcal{X}) + \tanh(\alpha h) \cdot \tanh(\alpha t)}. \tag{3.265d}$$

All of the previously studied transversely magnetized structures can be obtained as limiting cases of (3.265a)–(3.265d) by tending either h or t or both to zero or infinity. As proved in the previous sections, these are magneto-static forward volume waves (MSFVW).

3.27   LONGITUDINALLY MAGNETIZED SHIELDED DIELECTRIC FERRITE LAYERS

This structure may result from that of Figure 3.29 by placing a metallic screen on top of it at a distance h, as shown in Figure 3.32.

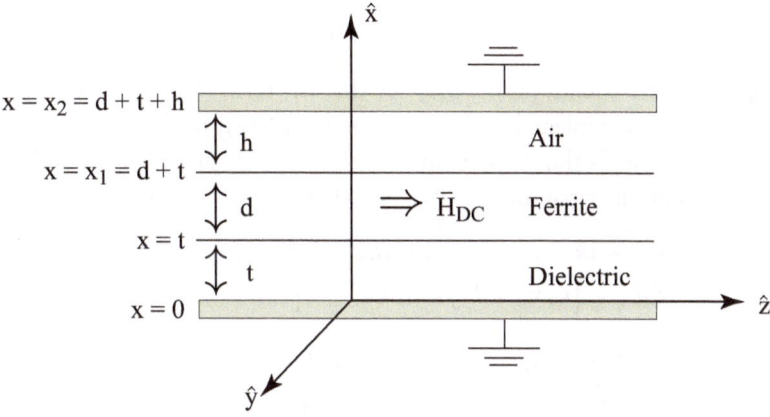

Figure 3.32: Shielded longitudinally magnetized dielectric–ferrite double layer.

In the present structure, the x-eigenfunction in the air region is as follows:

$$\psi_e = f_e(x)e^{jk_y y}e^{-jk_z z} \quad \text{where} \quad f_e(x) = A_e e^{-\alpha x} + B_e e^{+\alpha x}. \tag{3.266a}$$

This eigenfunction takes the appropriate form by asking for the normal magnetic induction $b_n = b_x$ to be zero on the metallic screen at $x = x_2$:

$$f_e(x) = 2A_e e^{-\alpha x_2} \cosh\left[\alpha\left(x - x_2\right)\right] = A'_e \cosh\left[\alpha\left(x - x_2\right)\right] \qquad (3.266b)$$

and

$$f'_e(x) = \alpha \cdot A'_e \cdot \sinh\left[\alpha\left(x - x_2\right)\right]. \qquad (3.266c)$$

3.27.1   MAGNETOSTATIC VOLUME MODES

The boundary conditions at the interface $(x = 0, t)$ remain as before, while new continuity conditions of ψ and b_x at $x = d + t$ should be enforced. These yield:

$$\begin{aligned}
A_x \cdot &\left[(1 + \mathcal{X})k_{xi} - \left[k_1 k_y - \alpha \cdot \tanh(\alpha h)\right]\tan\left[k_{xi}(d + t)\right]\right] \\
&= B_x \cdot \left[(1 + \mathcal{X})k_{xi}\tan\left[k_{xi}(d + t)\right] + \left[k_1 k_y - \alpha \cdot \tanh(\alpha h)\right]\right].
\end{aligned} \qquad (3.267)$$

Inspecting the above equation, one may see that the term $(k_1 k_y - \alpha)$ of the open structure occurring in Eq. (3.240a) is replaced by $[(k_1 k_y - \alpha \cdot \tanh(\alpha h)]$ in the above shielded case. This substitution in (3.240b) yields the dispersion equation, which can be also obtained by combining (3.267) with the boundary conditions at $x = t$. Either way, the result is:

$$\begin{aligned}
&\left.\frac{(1 + \mathcal{X})k_{xi} - \left[k_1 k_y - \alpha \cdot \tanh(\alpha h)\right]\tan\left[k_{xi}(d + t)\right]}{(1 + \mathcal{X})k_{xi} - \left[k_1 k_y - k_{xd} \cdot \tanh\left(k_{xd}t\right)\right]\tan\left(k_{xi}t\right)}\right|_{k_{xd}=\alpha} \\
&= \left.\frac{(1 + \mathcal{X})k_{xi}\tan\left[k_{xi}(d + t)\right] + \left[k_1 k_y - \alpha \cdot \tanh(\alpha h)\right]}{(1 + \mathcal{X})k_{xi}\tan\left(k_{xi}t\right) + \left[k_1 k_y - k_{xd} \cdot \tanh\left(k_{xd}t\right)\right]}\right|_{k_{xd}=\alpha}.
\end{aligned} \qquad (3.268)$$

One may observe in (3.268) the symmetry in the presence of the two metallic ground planes at distances $\Delta x = -t$ and $\Delta x = h$. The above dispersion equation can be simplified for the practically important case of pure propagation in the z-direction. Thus, (3.268) is reduced by setting $k_y = 0$ and using (3.222), as:

$$\tan\left(\frac{k_z d}{\sqrt{-(1 + \mathcal{X})}}\right) = \sqrt{-(1 + \mathcal{X})} \cdot \frac{\tanh\left(k_z h\right) - \tanh\left(k_z t\right)}{(1 + \mathcal{X}) - \tanh\left(k_z h\right) \cdot \tanh\left(k_z t\right)}. \qquad (3.269)$$

The previously discussed dispersion equations may result from (3.269) as limiting cases when h or t or both tend either to zero or infinity. It is interesting to examine the case when $h \to 0$, that is, when the top conductor is directly placed on top of the ferrite. This structure corresponds to a parallel plane waveguide loaded with dielectric-ferrite layers where (3.269) reduces:

$$\tan\left(\frac{k_z d}{\sqrt{-(1 + \mathcal{X})}}\right) = \frac{\tanh\left(k_z t\right)}{\sqrt{-(1 + \mathcal{X})}}. \qquad (3.270)$$

The magnetostatic volume modes dispersion equation of a parallel plate waveguide loaded only with ferrite can be readily obtained from (3.270) by setting $t = 0$.

3.27.2   MAGNETOSTATIC SURFACE MODES $(1 + \mathcal{X}) > 0$

Imposing the additional boundary conditions at $x = x_1$, x_2 we end up with the following dispersion equation:

$$
\left. \frac{\left\{(1 + \mathcal{X})\alpha_i - \left[k_1 k_y - \alpha \cdot \tanh(\alpha h)\right]\right\} e^{\alpha_i d}}{(1 + \mathcal{X})\alpha_i - \left[k_1 k_y - k_{xd} \cdot \tanh(k_{xd} t)\right]} \right|_{k_{xd} = \alpha}
$$
$$
= \left. \frac{\left\{(1 + \mathcal{X})\alpha_i + \left[k_1 k_y - \alpha \cdot \tanh(\alpha h)\right]\right\} e^{-\alpha_i d}}{(1 + \mathcal{X})\alpha_i + \left[k_1 k_y - k_{xd} \cdot \tanh(k_{xd} t)\right]} \right|_{k_{xd} = \alpha}. \tag{3.271}
$$

This is similar to (3.242) with only an additional $\tanh(\alpha h)$ term. For the practical case of pure propagation in the y-direction or $k_z = 0$, the above dispersion equation reduces to:

$$
e^{-2|k_y|d} = \frac{(1 + \mathcal{X}) - k_1 \cdot S + \tanh\left(|k_y| h\right)}{(1 + \mathcal{X}) + k_1 \cdot S - \tanh\left(|k_y| h\right)} \cdot \frac{(1 + \mathcal{X}) + k_1 \cdot S - \tanh\left(|k_y| t\right)}{(1 + \mathcal{X}) - k_1 \cdot S + \tanh\left(|k_y| t\right)}. \tag{3.272}
$$

The above resembles (3.244) when $h \to 0$. The practical, useful case of finite slabs when the layers are truncated in the z-direction can be characterized by inserting expressions (3.250) into (3.272). The details are left as a problem for an interested reader.

3.28   MAGNETIZED AND SPIN WAVES IN FERRITE SLAB WITH LOSSES

The general form of the magnetization equation considering ferrite losses was given through Eqs. (2.28)—(2.30). The same general form applies to magnetostatic and spin waves as well. Numerous authors, e.g., Gurevich [52] and Vyzulin et al. [53], studied the effect of losses on these waves. As with the general case, which is also valid for electromagnetic waves, losses can be included by substituting the resonant frequency $\omega_i = \mu_0 \gamma H_i$ by a complex value [52]:

$$
\omega_i \; \leftarrow \; \omega_i + j \omega \alpha_s \tag{3.273}
$$

where α_s is the magnetostatic or spin-waves damping factor, a dimensionless parameter. Once again, the damping factor can be obtained from the measurements of the magnetostatic or spin-wave resonance:

$$
\alpha_s = \frac{\mu_0 \gamma \Delta H_s}{2 \omega_i} \quad \text{or} \quad \omega \alpha_s = \frac{1}{2} \frac{\omega}{\omega_i} \mu_0 \gamma \Delta H_s = \frac{1}{2} \omega \cdot \frac{\Delta H_s}{\omega_i} \tag{3.274}
$$

where ΔH_s is the magnetostatic or spin-wave resonance linewidth. This is measured as the two frequency or H_i points where the resonance response drops to half of its maximum, just as shown in Figure 2.13. At this point, be careful to understand that the magnetostatic or spin waves linewidth ΔH_s is different than that of the electromagnetic modes (ΔH), since it is evaluated from measurements at the corresponding mode resonances. For this reason, tables characterizing ferrite materials provide different linewidths for electromagnetic modes and spin waves.

In view of the substitution of (3.273) into (3.271), the magnetic susceptibility elements $\mathcal{X}$ and k_1 become complex as $\mathcal{X} = \mathcal{X}' - j\mathcal{X}''$ and $k_1 = k_1' - jk_1''$. Hence, all the resulting wavenumbers also become complex $k = k' - jk''$ and the propagation dependence is modified as $e^{-jkw} = e^{-k''w}e^{-jk'w}$ with $w = x, y, z$. That is, the wave becomes inhomogeneous with attenuation constant k''. In turn, one may go through the previous analysis and make the above substitutions to account for ferrite losses.

Some indicative examples of such modified expression can be found in [53]. Note that ferrite losses have important effects in the frequency range around the gyromagnetic resonance $\omega \rightarrow \omega_i$, where the susceptibility becomes finite. This, in turn, drastically restricts the number of possibly excited modes. However, the most important effect of losses is the occurrence of new branches in the magnetostatic wave spectrum.

3.29   MAGNETOSTATIC WAVE SPECTRUM IN THE PRESENCE OF LOSSES

Vyzulin et al. [53] first considered the effect of losses on the magnetostatic wave spectrum. They observed new branches in their dispersion curves when losses are considered. In order to clarify this effect, recall that for the lossless case ($\mathcal{X}$ = real), the frequency range of surface and volume modes was based on the wavenumber relation (3.135) and the sign of $(1 + \mathcal{X})$ to yield the ranges of Eq. (3.135). Let us recall (3.135) inside the ferrite region I as:

$$(1 + \mathcal{X}) \cdot \left[k_{xi}^2 + k_{yi}^2 \right] + k_{zi}^2 = 0. \tag{3.275}$$

When $\mathcal{X}$ is real, and the volume-wave modes require that $k_{xi}^2, \ k_{yi}^2, \ k_{zi}^2 > 0$, the only possible solution occurs when $(1 + \mathcal{X}) < 0$. Respectively, surface waves require $k_{xi}^2 < 0$, and they may occur only when $(1 + \mathcal{X}) > 0$. On the contrary, when losses are included, $\mathcal{X}$ becomes complex and the situation is modified. The volume modes require all three k_{xi}, k_{yi}, and k_{zi} to have a significant real part and a small imaginary one. This should hold for k_{yi} and k_{zi} for surface waves, while k_{xi} should have a large imaginary and small real part.

3.30   REFERENCES

[1] C. L. Hogan, The elements of nonreciprocal microwave devices, *IRE Proc.*, vol. 44, pp. 1345–1368, October 1956. DOI: 10.1109/jrproc.1956.274979. 73, 74, 76, 77

[2] C. Kittel, On the theory of ferromagnetic resonance absorption, *Physical Review*, vol. 73, pp. 155–161, 1948. DOI: 10.1103/physrev.73.155. 73, 76, 101

[3] D. M. Pozar, *Microwave Engineering*, 4th ed., Wiley, NY, 2016. 74, 75, 76, 112, 133, 141

[4] B. Fuller, *Microwave Ferrites Engineering*, 2nd ed., Wiley, NY, 1998. 74, 75, 78, 79, 81

[5] R. A. Waldron, *Ferrites: An Introduction for Microwave Engineers*, Van Nostrand, NY, 1961. 76, 77, 78, 79

[6] C. Kittel, Interpretation of anomalous larmor frequencies in ferromagnetic resonance experiment, *Physical Review*, vol. 71, pp. 270–1, 1947. DOI: 10.1103/physrev.71.270.2. 76, 81

[7] V. Denysenkov, Broadband ferromagnetic resonance spectrometer: Instrument and application, Ph.D. Thesis, Royal Institute of Technology, Stockholm, 2003. DOI: 10.1063/1.1581395. 78, 79

[8] C. Kittel, *Introduction to Solid State Physics*, John Wiley, NY, 1996. DOI: 10.1063/1.3060399. 78, 79, 97

[9] L. R. Walker, Ferromagnetic resonance: Line structures; resonant modes of ferromagnetic spheroids, *Journal of Applied Physics*, vol. 29, no. 3, pp. 318–323, March 1958. (Note: This is an important survey article for spin waves and magnetostatic modes). 79, 100, 117

[10] L. R. Walker, Magnetostatic modes in ferromagnetic resonance, *Physical Review*, vol. 105, no. 2, pp. 390–399, January 1957. DOI: 10.1103/physrev.105.390. 79, 80, 97, 98, 100

[11] M. Pardavi-Horvath, B. Keszei, J. Vandlik, and R. D. McMichael, Ferromagnetic resonance spectra of thick yttrium-iron-garnet films, *Journal of Applied Physics*, vol. 87, no. 9, pp. 4969–4971, May 2000. DOI: 10.1063/1.373218.

[12] W. K. Hiebert, Experimental micromagnetic dynamics: Ultrafast magnetization reversal using time resolved scanning kerr effect microscopy, Ph.D. Thesis, University of Alberta, Dept. of Physics, Canada, 2001. xv, 80, 81, 82, 84, 85, 86, 88, 95, 114, 124, 125, 127, 128, 129, 131

[13] R. W. Damon and J. R. Eshbach, Magnetostatic modes of a ferromagnet slab, *Journal of Physics and Chemistry of Solids*, vol. 19, no. 3-4, pp. 308–320, 1961. DOI: 10.1016/0022-3697(61)90041-5. xv, 80, 101, 115, 117, 119, 120, 121, 122, 126, 127, 129, 130, 131, 145, 150, 151

[14] G. E. Ballentine, Comparison of time resolved micromagneticc dynamics experiments on $Ni_{80}Fe_{20}$ and Landau–Lifshitz–Gilbert micromagnetic simulation, Ph.D. Thesis, Dept. of Physics, Univ. of Alberta, Canada, 2002. 81

[15] R. Urban, Electron tunnelling and spin dynamics and transport in crystalline magnetic multilayers, Ph.D. Thesis, Dept. of Physics, Simon Fraser University, September 2003. 81

[16] C. Jozsa, Optical detection of the magnetization precession, Ph.D. Thesis, Technical University of Eindhoven, 2006. xvi, 83, 85, 87, 135, 136

[17] F. R. Morgenthaler, An overview of electromagnetic and spin angular momentum mechanical waves in ferrite media, *Proc. of the IEEE*, vol. 76, pp. 138–150, February 1988. DOI: 10.1109/5.4390. 85, 86, 102, 114, 115

[18] W. P. Wolf, Ferrimagnetism, *Reports on Progress in Physics*, vol. 24, no. 1, pp. 212–303, 1961. DOI: 10.1088/0034-4885/24/1/306. xv, 84, 86, 95, 96

[19] T. Holstein and H. Primakoff, Field dependence of the intrinsic domain magnetization of a ferromagnet, *Physical Review*, vol. 58, pp. 1098–1113, 1940. DOI: 10.1103/physrev.58.1098. 86

[20] C. E. Patton, Magnetic excitations in solids, *Physics Reports (Review Section of Physics Letters)*, vol. 103, no. 5, pp. 251–315, Elsevier Pub., 1984. DOI: 10.1016/0370-1573(84)90023-1. xv, 86, 88, 95, 96

[21] M. Sparks, *Ferromagnetic Relaxation Theory*, McGraw-Hill, NY, 1964. 86, 137, 139, 141

[22] E. Schlomann, Generation of spin waves in nonuniform magnetic fields. I. Conversion of electromagnetic power into spin wave power and vice-versa, *Journal of Applied Physics*, vol. 35, no. 1, pp. 159–166, January 1964. DOI: 10.1063/1.1713058. 86

[23] H. Suhl, The theory of ferromagnetic resonance at high signal powers, *Journal of Physics and Chemistry of Solids*, vol. 1, pp. 209–227, Pergamon Presss, 1957. DOI: 10.1016/0022-3697(57)90010-0. 87, 89, 96

[24] O. Heinonen, Computational applied magnetics, *Seagate Technologies*, Boulder MUMAG-03 Summer School, July 2003. xv, 82, 84, 89, 90, 91, 101

[25] C. Herring and C. Kittel, On the theory of spin waves in ferromagnetic media, *Physical Review*, vol. 81, no. 5, pp. 869–880, March 1951. DOI: 10.1103/PhysRev.81.869. 82, 87, 89, 92, 94

[26] P. B. Visscher, D. M. Apalkov, and X. Feng, Understanding magnetic switching: Spin wave visualization, Department of Physics, University of Alabama, 2001. 88

[27] J. D. Jackson, *Classical Electrodynamics*, 3rd ed., Wiley, NY, 1998. DOI: 10.1063/1.3057859. 89, 147

[28] C. R. Buffler, Ferromagnetic resonance near the upper limit of the spin wave manifold, *Journal of Applied Physics*, vol. 30, 172(S). DOI: 10.1063/1.2185872. xv, 92, 96

[29] C. Kittel, Excitation of spin waves in a ferromagnet by a uniform RF field, *Physical Review*, vol. 110, p. 1295, 1958. DOI: 10.1103/PhysRev.110.1295. 97

[30] G. A. Vugalter and I. A. Gilinskii, Magnetostatic waves: A review, *Radiofizika*, vol. 32, no. 10, pp. 1187–1220, October 1989. English Translation: Plenum Publishing, 1996. DOI: 10.1007/bf01053398. 85, 97, 100, 101

[31] P. W. Anderson and H. Suhl, Instability in the motion of ferromagnets at high microwave power levels, *Physical Review*, vol. 100, pp. 1788–9, 1955. DOI: 10.1103/physrev.100.1788. 97, 98

[32] H. Suhl, The nonlinear behavior of ferrites at high microwave signal levels, *Proc. of the Institute of Radio Engineers*, vol. 44, no. 10, pp. 1270–1284, 1956. DOI: 10.1109/jrproc.1956.274950. 97

[33] H. Suhl, *Lectures on Magnetism*, University of Colorado, Boulder, CO, 2006. 97, 98

[34] R. Loudon and P. Pincus, Effect of dipolar fields on the antiferromagnetic spin wave spectrum, *Physical Review*, vol. 132, no. 2, pp. 673–678, October 1963. DOI: 10.1103/physrev.132.673. 99

[35] F. Keffer and C. Kittel, Theory of antiferromagnetic resonance, *Physical Review*, vol. 85, no. 2, pp. 329–337, January 1952. DOI: 10.1103/physrev.85.329. 99

[36] J. R. MacDonald, Ferromagnetic resonance and the internal field in ferromagnetic materials, *Physical Review*, vol. 103, p. 280, 1956. DOI: 10.1088/0370-1298/64/11/302. 101

[37] R. W. Damon and H. Van De Vaart, Propagation of magnetostatic spin waves at microwave frequencies in a normally magnetized disk, *Journal of Applied Physics*, vol. 36, no. 11, pp. 3453–3459, November 1965. DOI: 10.1063/1.1703018. 101, 137

[38] B. A. Kalinikos, Excitations of propagating spin waves in ferromagnetic films, *IEE Proc., Part H*, vol. 127, pp. 4–10, 1980. DOI: 10.1049/ip-h-1.1980.0002. 81, 124, 127

[39] M. Abramovitz and I. A. Stegun, *Handbook of Mathematical Functions*, Dover Publishing, NY, 1965. 127, 141, 142, 144, 154

[40] M. J. Hurben and C. E. Patton, Theory of magnetostatic waves for in-plane magnetized anisotropic films, *Journal of Magnetism and Magnetic Materials*, vol. 163, pp. 39–69, 1996. DOI: 10.1016/s0304-8853(96)00294-6. 131

[41] M. J. Hurben and C. E. Patton, Theory of magnetostatic waves for in-plane magnetized isotropic films, *Journal of Magnetism and Magnetic Materials*, vol. 139, pp. 236–291, 1995. DOI: 10.1016/0304-8853(95)90006-3. xv, 131, 132, 135

[42] M. Sparks, Magnetostatic modes in an infinite circular disk, *Solid State Communications*, vol. 8, pp. 731–733, 1970. DOI: 10.1016/0038-1098(70)90419-9. 137, 141

[43] A. Z. Maksymowicz, The magnetic resonance modes and their intensities for a thin slab, *Thin Solid Films*, vol. 42, pp. 245–253, Elsevier, 1977. DOI: 10.1016/0040-6090(77)90423-0. 140

[44] P. C. Fletcher and C. Kittel, Considerations the propagation and generation of magnetostatic waves and spin waves, *Physical Review*, vol. 120, no. 6, pp. 2005–6, December 1960. DOI: 10.1103/physrev.120.2004. 142, 144

[45] R. I. Joseph and Schlomann, Theory of magnetostatic modes in long axially magnetized cylinders, *Journal of Applied Physics*, vol. 32, 1001, 1961. DOI: 10.1063/1.1736149. 142, 143, 144

[46] E. O. Kamenetskii, Handedness of magnetic dipolar modes in ferrite disks, *Journal of Magnetism and Magnetic Materials*, vol. 302, no. 1, pp. 137–155, July 2006. DOI: 10.1016/j.jmmm.2005.09.001. 142, 143

[47] W. L. Bongianni, Magnetostatic propagation in dielectric layered structure, *Journal of Applied Physics*, vol. 43, no. 6, pp. 2541–8, June 1972. DOI: 10.1063/1.1661557. 145, 151, 155

[48] S. R. Seshadri, Surface magnetostatic modes of a ferrite slab, *IEEE Proc.*, vol. 58, pp. 506–7, March 1970. DOI: 10.1109/proc.1970.7680. 145, 149, 150, 151, 156

[49] T. W. O'Keefe and R. W. Patterson, Magnetostatic surface wave propagation in finite samples, *Journal of Applied Physics*, vol. 49, pp. 4886–4895, 1978. DOI: 10.1063/1.325522. 155, 156, 157

[50] W. S. Ishak, Magnetostatic wave technology: A review, *IEEE Proc.*, vol. 76, pp. 171–187, February 1988. DOI: 10.1109/5.4393. 151, 156

[51] J. P. Parekh, K. W. Chang, and H.S. Tuan, Propagation characteristics of magnetostatic waves, *Circuits Systems Signal Process*, vol. 4, no. 1–2, pp. 9–39, 1985. DOI: 10.1007/bf01600071. 161

[52] A. G. Gurevich, Magnetic resonance in ferrites and antiferromagnetics, *Science*, p. 39, 1973. 165

[53] S. A. Vyzulin, A. E. Rosenson, and S. A. Shekh, The magnetostatic waves in ferrite film with losses, *IEEE Transactions on Microwave Theory and Techniques*, vol. MTT-41, pp. 1070–3, June/July 1973. DOI: 10.1109/22.238528. 165, 166

CHAPTER 4

Multiferroics: Ferrite–Ferroelectric Composites

4.1 INTRODUCTION

Tunable ferrite microwave structures exhibit low-loss electromagnetic wave propagation, broadband frequency operation, and tunability in a high dynamic range. In particular, devices based on ferrite films or ferrite posts (mainly made of YIG) are less bulky than the classical ones. However, their key drawback stems from the necessity of electronically controllable static magnetic field H_{DC}. Despite all progress made, magnets remain bulky and heavy. Thus, the overall structure is unable to be integrated in current microwave devices. Ferroelectrics arrived as a very promising solution. Their relative permittivity can be tuned by a static electric field E_{DC}. With extraordinary materials like barium strontium titanate ($Ba_{1-x}Sr_xTiO_3$, BSTO), they are feasible at room temperatures. Tunable and exceptionally non-reciprocal nanostructured devices could be integrated in a chip, bringing a revolution in microwave front ends and phased arrays. Unfortunately, despite the huge effort and extreme financial support, the practical results were negligible. There were very high losses in ferroelectric media even when incorporated in resonant structures as cavities.

The logical question arising from the above rationale is whether it is possible to invent any media exhibiting the low loss, broadband operation, and high dynamic range like ferrites (especially films) in conjunction with the exceptionally attractive electric field (DC voltage) tunability of ferroelectrics? Although it once seemed impossible, multiferroics indeed exhibit all these extraordinary characteristics [1]. The first invented natural multiferroics exhibited the desired characteristics at very low temperatures with poor tunability, equivalent to a very weak DC magnetic field. However, scientists soon realized that the answer lie in double- or multi-phase or ferrite-ferroelectric composites (magnetoelectric composites) which yield to "giant" magnetoelectric response. These media are so promising that one may expect to fulfill all the requirements for nanostructured microwave devices and, in turn, RF front ends as well as phased array control units to be integrated onto a chip. Beyond that, the availability of on-chip non-reciprocal microwave properties may enable unexplored features. Besides the strictly MW engineering applications, multiferroics may support sensors of electric or magnetic-field sensors of mechanical stress though elastic interactions and heterogeneous read/write devices [1]. Regarding the latter feature, a completely new scientific field of "spintronics" is developing [1, 2]. That is, the ability of an indirect switch of ferrite magnetization by a DC voltage leads to the switch of its electron spins. Hence, spin switching can be exploited and is expected to result in huge and very fast multi-state memories, or digital storage in general. Scientists' ambition goes far beyond this feature, to the exploitation of spin switching in transistors operating at microwave frequencies. It is to these possible evolutions that "spintronics" are expected to respond.

Although the above described multiferroic composites are expected to bring a revolution in MW front ends and phased arrays, their possibilities for exploitation are yet unexplored. Researchers are invited to enter this scientific quest. A readily seen evolution, including all the above positive factors, would address tunable single antenna elements, microwave filters, and resonators incorporated to voltage-controlled oscillators. Let us focus on tunable single-printed antenna elements. The voltage control of its substrate constitutive relation $\mu_r(V_{DC})$, $\varepsilon_r(V_{DC})$ may enable both its resonant frequency control $f_{res}(V_{DC})$ as well as its radiating beam maximum orientation. Antennas printed on magnetized ferrite substrate partly offer these features already. The ferrite may be incorporated as film or post to reduce its weight and size as well as the controlling DC magnetic field requirements. The radiator is miniaturized as its size is inversely proportional to $1/\sqrt{\mu_r \varepsilon_r}$ and its characteristics are tuned based on $\mu_r(H_{DC})$. Note that ferrites like YIG exhibit a high dielectric constant of the order of $\varepsilon_r = 12$ to 14, while its tunable μ_r varies between 0.5–1.0. Besides the resulting bulky structure due to the electromagnet, this type of antenna has a very narrow frequency bandwidth due to its high-Q factor. The usage of a multiferroic composite film in its substrate first makes its tunability very conveniently controlled by a DC voltage, but also it can substantially increase its bandwidth, even resulting in a broadband element. This latter feature stems from the possible simultaneous control of both $\mu_r(V_{DC})$ and $\varepsilon_r(V_{DC})$. Explicitly, the patch antenna quality factor is proportional to [3]:

$$Q \approx \frac{4}{\sqrt{2}} \sqrt{\frac{\mu_r}{\varepsilon_r}} \frac{d}{\lambda_{res}} = \frac{f_{res}}{BW}, \quad \text{where } d = \text{substrate thickness.} \tag{4.1}$$

Hence, the miniaturization can be achieved by a factor $1/\sqrt{\mu_r \varepsilon_r}$, while the tuning can be done in such a way as to retain $\mu_r(V_{DC}) = \varepsilon_r(V_{DC})$ to exhibit the lowest Q or the maximum possible bandwidth [3].

4.2   MULTIFERROIC PROPERTIES

Multiferroic media exhibit at least two coexistent orders of ferroelectric-ferroelastic and ferromagnetic media [1]. The magneto-electric effect (ME_H) refers to the appearance of an electric polarization $\overline{p}$ due to the application of magnetic field H as $(\overline{P} = \alpha_m \overline{H})$. Correspondingly, the converse magneto-electric yields a magnetization $\overline{M}$ when an electric field $\overline{E}$ is applied as $(\overline{M} = a_e \overline{E})$, (see [1–4] and the original references therein). Natural or single-phase multiferroics are very rare and their magneto-electric response is either too weak or occur at very low temperatures, making them impractical [1]. On the contrary, multiferroic composites comprising both ferroelectric and ferrimagnetic or ferromagnetic phases may yield "giant" magnetoelectric coupling (as $10^2 \left(\frac{V}{cm} \right)$Oe) at or above room temperatures. This readily offers practical applications. Note that the magnetoelectric response of these composites is up to several orders of magnitude higher than that of single-phase multiferroics.

The magneto-electric (ME) effect in a two-phase composite comes from the cross interaction between different orderings of its phases. For this purpose, one of the two phases must exhibit double orders. Usually, the one phase exhibits both ferroelectric and piezoelectric properties, while the second is a magnetic phase. Neither of the two constituents exhibit a magnetoelectric phase, but this appears in their composite through a cross coupling. That is, an electric phenomenon is coupled to a magnetic one through elastic interaction.

The interaction scheme can be interpreted as [1]:

Magnetic field $\longrightarrow$ mechanical phenomena $\longrightarrow$ electric polarization:

$$ ME_H \text{ effect } = \frac{\text{magnetic } (H)}{\text{mechanical } \sigma} \times \frac{\text{mechanical } (\sigma)}{\text{electric } (E)}. \tag{4.2} $$

Electric field $\longrightarrow$ mechanical phenomena $\longrightarrow$ magnetic polarization:

$$ ME_E \text{ effect } = \frac{\text{electric } (E)}{\text{mechanical } \sigma} \times \frac{\text{mechanical } (\sigma)}{\text{magnetic } (H)}. \tag{4.3} $$

For the ME_E effect in (4.3) scheme, an electric field is applied to the composite and mechanical oscillation, or shape change is caused to the piezoelectric/ferroelectric phase. The resulting strain is passed to the magnetic phase, making a magnetic polarization due to its magnetostriction. Conversely, for the ME_H effect in (4.2) scheme, a magnetic field applied to the composite causes a magneto-strictive shape change to the magnetic phase. This mechanical strain is transferred to the piezoelectric/ferroelectric phase, resulting in electric polarization through piezoelectricity.

A general illustration of all electric, mechanical-elastic, and magnetic interactions is given in Figure 4.1, adapted from [3] and the original references therein. Observe from Figure 4.1a that single-phase multiferroics exhibit both ferroelectric and ferro-or ferri-magnetic properties, while these two orders identify another distinctive magnetoelectric category. All possible interactions are depicted in Figure 4.1b along with the corresponding constitutive parameters. That is, we have the classical direct polarization $\overline{E}$ to $\overline{P}$ through electric susceptibility χ_e: the magnetic one $\overline{H}$ to $\overline{M}$ through magnetic susceptibility χ_m and the elastic σ to $\overline{\varepsilon} = \overline{S}$ through the stiffness (C). However, the same Figure 4.1b depicts the indirect interactions as ME_E by $\overline{\varepsilon}$ to σ through piezoelectricity and, in turn, σ to $\overline{H}$ through magnetoelasticity.

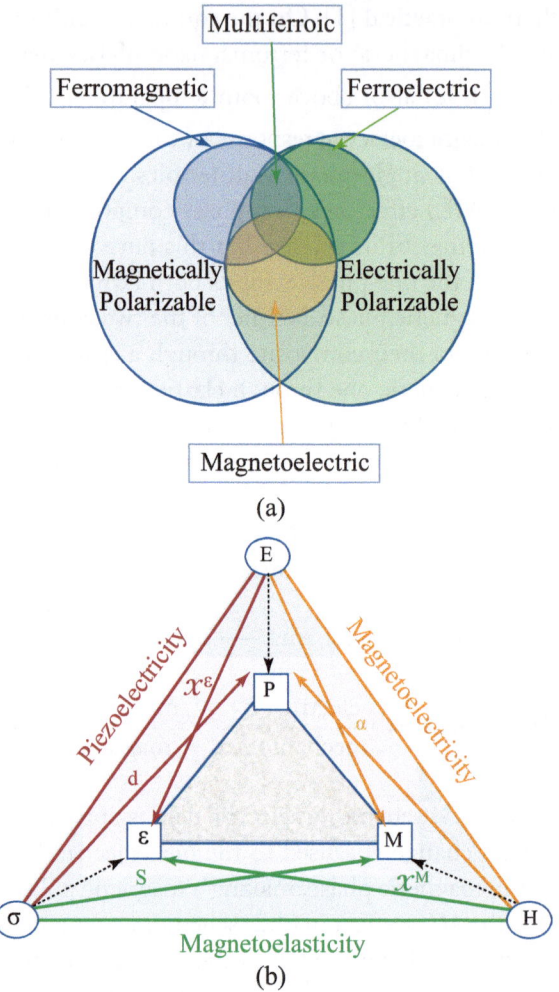

Figure 4.1: (a) Relationships and overlapping of ferro-properties in complex media. (b) All possible interactions in multiferroic and magneto-electric media [3].

4.3 TOPOLOGIES—CONNECTIVITY AT TWO–PHASE COMPOSITES

Multiferroic composites are built up by embedding one phase (usually the magnetic) as inclusions in the matrix of the other phase (usually the ferroelectric/piezoelectric), as shown in Figure 4.2. These are defined by different connectivity schemes denoted by two numbers classified by Newnham et al [5]. The first number denotes the connectivity of one-phase particle inclusions embedded in the matrix of the other phase, which is denoted by the second number. The often built composites and their connectivity are illustrated in Figure 4.2 as: (a) spherical particles, (b) laminate composite, and (c) fiber/rod inclusions. Usually, materials such as $BaTiO_3$, PZT (Terfenol-D ceramic), piezoelectric polymer [PCVDF_TrFE], and $Pb(MgNb)O_3 – PbTiO_3$ are utilized for the matrix ferroelectric/piezoelectric phase, while ferrites—often nickel (Ni), lithium (Li) or YIG for the magnetic phase—comprise the particles inclusions [1].

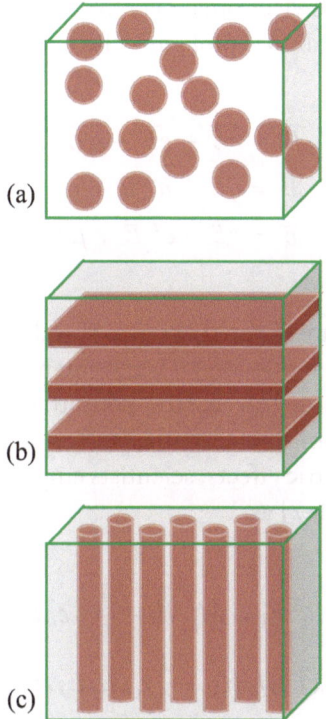

(a)

(b)

(c)

Figure 4.2: Three usual multiferroic composites: (a) spherical particles, (b) laminate composite, and (c) fiber/rod inclusions.

The laminate composite (Figure 4.2b) is also referred to as layered multiferroic heterostructure of alternating layers of ferroelectric and magnetic thin films. In [6], there were magnetite/lead-zirconium-titanate (Fe_3O_4/P2T) and lead-magnesium niobate-lead titanate (Fe_3O_4/PMN-PT) multiferroic heterostructures employed. Also reported was an electrostatically tunable ferromagnetic resonance (FMR) field shift of up to 600 Oe, corresponding to a microwave *ME* coefficient of 670 Oe/(kV/cm).

4.4   MULTIFERROICS CONSTITUTIVE RELATIONS

The design of any microwave device is based on an initial estimation of the geometry-topology. An accurate simulation up to its final layout form, ready for fabrication, follows this. Devices involving ordinary media are based on electromagnetic simulations requiring the constitutive relations between electric and magnetic-field displacement/induction and the corresponding field intensities. However, when multiferroics are incorporated, the mechanical/elasticity problem must be solved in parallel with the electromagnetic one. In other words, the simulation corresponds to the solution of a multi-physics problem. The involved constitutive relations read [1]:

$$\overline{D} = e\overline{S} + \varepsilon\overline{E} + a\overline{H} \tag{4.4a}$$

$$\overline{B} = q\overline{S} + a^T\overline{E} + \mu\overline{H} \tag{4.4b}$$

$$\overline{\sigma} = c\overline{S} - e^T\overline{E} + \mu\overline{H}. \tag{4.4c}$$

The vector-field quantities are devoted as stress $\overline{\sigma}$ and strain $\overline{S}$, electric field displacement ($\overline{D}$), intensity ($\overline{E}$), magnetic field induction ($\overline{B}$), and intensity ($\overline{H}$). Correspondingly, the constitutive parameters are the stiffness (c), electric permittivity (ε), magnetic permeability (μ), piezoelectric coefficient (e), piezomagnetic coefficient (q), and ME coefficient (a). For a better understanding of the magnetoelectric effects, scientists employ the expansion of the free-energy functional system-Hamiltonian [4–7] as:

$$F\left(\overline{E}, \overline{H}\right) = F_o - P_i^S E_i - M_i^S H_i - \frac{1}{2}\varepsilon_o\varepsilon_{ij}E_i E_j - \frac{1}{2}\mu_o\mu_{ij}H_i H_j$$
$$- \frac{1}{2}\beta_{ijk}E_i H_j H_k - \frac{1}{2}\gamma_{ijk}H_i E_j H_k \ldots \tag{4.5}$$

The differentiation of the energy functional (Hamiltonian) with respect to the applied microwave electric field intensity yields the electric polarization:

$$P_i\left(\overline{E}, \overline{H}\right) = -\frac{\partial F}{\partial E_i} = P_i^S + \varepsilon_o\varepsilon_{ij}E_j + a_{ij}H_i + \frac{1}{2}\beta_{ijk}H_j H_k + \gamma_{ijk}H_i E_j \ldots \tag{4.6a}$$

Likewise, the differentiation with respect to the applied magnetic-field intensity (H_i) yields the magnetization:

$$M_i\left(\overline{E}, \overline{H}\right) = -\frac{\partial F}{\partial H_i} = M_i^S + \mu_o\mu_{ij}H_j + a_{ij}E_i + \beta_{ijk}E_i H_j + \frac{1}{2}\gamma_{ijk}E_j E_k \ldots \tag{4.6b}$$

where $a = a_{ij}$ is the linear magneto-electric coefficient, while β_{ijk} and γ_{ijk} represents the higher-order magneto-electric effects which are usually negligibly small with respect to a_{ij}.

An interesting example of multiferroics adopted from [1] and the original references therein is presented in Figure 4.3. This refers to three different bilayers laminate composite: (i) nickel ferrite ($NiFe_2O_4$)/PZT, (ii) lithium ferrite/PZT, and (iii) yitrium iron garnet (YIG)/PZT. For practical applications, these composites are subject to a constant static magnetic field H_{DC} (e.g., provided by a permanent magnet) to be selected with the aid of diagrams like that of Figure 4.3, while their ferromagnetic resonance (FMR) will be tuned by a static electric field. The above figure presents the magnetic susceptibility χ_m vs. the static magnetic field (H_{DC}) without an external DC electric field (curves green) and with a strong electric field $E_{DC} = 300\,kV/cm$ (curves red). The stronger tunability is observed for the nickel ferrite (NFO/PZT), and a useful tunability is observed for the lithium ferrite/PZT, while a negligible tunability is observed for YIG/PZT bilayer.

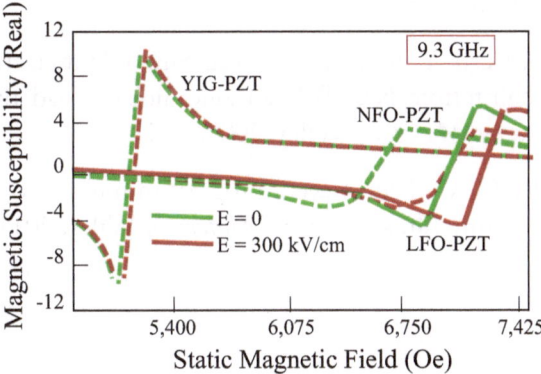

Figure 4.3: Theoretical electric field magnetic tunability for three bilayers: (i) nickel ferrite ($NiFe_2O_4$)/PZT, (ii) lithium ferrite/PZT, and (iii) yitrium iron garnet (YIG)/PZT. E_{DC} can be obtained by a DC voltage of 30 V for a thin film composite with thickness of $1\,\mu m$ [1].

Further details regarding simulation and modeling of multiferroics can be found in the excellent review papers [1, 4] and the original references therein.

Multiferroics are very promising tuning materials in antenna and microwave applications and there is much research work to be done in the near future.

4.5   REFERENCES

[1] Ce-Wen Nan, M. I. Bichurin, S. Dong, D. Viehland, and G. Srinivasan, Multiferroic magnetoelectric composites: Historical perspective, status, and future directions, *Journal of Applied Physics*, vol. 103, 031101, 2008. xvi, 171, 172, 173, 175, 176, 177

[2] S. Fusil, V. Garcia, and A. Barthélémy, Magnetoelectric devices for spintronics, *Annual Review of Materials Research*, vol. 44, pp. 91–116, 2014. First published online as a Review in Advance on March 21, 2014. DOI: 10.1146/annurev-matsci-070813-113315. 172

[3] O. Rybin and S. Shulga, Profile miniaturization and performance improvement of a rectangular patch antenna using magnetic metamaterial substrates, *International Journal of RF and Microwave*, Wiley, 2016. xvi, 172, 173, 174

[4] L. W. Martin, S. P. Crane, Y-H Chu, M. B. Holcomb, M. Gajek, M. Huijben, C-H Yang, N. Balke, and R. Ramesh, Multiferroics and magnetoelectrics: Thin films and nanostructures, *Journal of Physics: Condensed Matter*, 20, 2008, 434220. DOI: 10.1088/0953-8984/20/43/434220. 172, 176, 177

[5] R. E. Newnham, D. P. Skinner, and L. E. Cross, Connectivity and piezoelectric-pyroelectric composites, *Materials Research Bulletin*, vol. 13, no. 525, 1978. DOI: 10.1016/0025-5408(78)90161-7. 174

[6] J. Lou, M. Liu, D. Reed, Y. Ren, and Nian X. Sun, Giant electric field tuning of magnetism in novel multiferroic FeGaB/Lead zinc niobate–lead titanate (PZN-PT) heterostructures, *Advanced Materials*, Wiley, 2009. 175

[7] J. Ma, J. Hu, Z. Li, C.-W. Nan, Recent progress in multiferroic magnetoelectric composites: from bulk to thin films, *Advanced Materials*, Wiley, 2011. 176

C H A P T E R   5

Planar Transmission Lines

5.1   INTRODUCTION

The primary purpose served by tunable ferrite/ferroelectric materials is the miniaturization of microwave and millimeter wave devices. This, in turn, asks for technologies compatible with microwave integrated circuits (MMICs), which require the introduction of tunable materials in the form of thin films. Working toward this direction, one may start from simple printed circuits and move forward to thick films and finally thin films. It is also clear that the design of any microwave circuit presumes the knowledge of the corresponding transmission line characteristics, which actually gives the name to the adopted technology. These can be identified as the stripline, the ordinary microstrip, the suspended and inverted microstrip, the slotline, the fin-line, and the coplanar waveguides. However, examining even the simple microstrip fabrication process, we can see that it does not allow its realization on a single ferrite or ferroelectric substrate. For example, the metallic strip cannot be printed-developed on the hard ferrite layer. Hence, considering the fabrication details, the transmission lines involve layers or films of ferrite or ferroelectric materials in a multilayer structure. This is composed of a multilayer substrate and an additional single or multilayer superstrate. The transmission lines printed on a single dielectric substrate and, at most, with a single dielectric superstrate are given in a lot of textbooks, e.g., Pozar [1], Collin [2], Hoffman [3], and Gupta et al. [4]. However, multilayer dielectric substrates are not elaborated on enough in textbooks, even though they are well studied in the published research literature. Thus, in the following sections, the characteristics of printed transmission lines involving multilayer dielectrics will be given, focusing on the design approach. In turn, transmission lines printed on tunable substrates will be elaborated upon. Next, transmission lines composed of tunable layers and multi-dielectric layers will be presented focusing on the dispersion curves and their characteristic impedance.

5.2   MULTILAYER MICROSTRIP LINES

The characteristics of multilayer microstrip lines have been the subject of much investigation since the early 1970s. The first attempts toward this analysis employed numerical methods such as the variational technique of Yamashita [5] or the numerical evaluation of the Green's function given by Farrar and Adams [6]. The above techniques as well as other numerical methods generally provide the desired accuracy. However, there is a need to have physical insight and, most importantly, a convenient engineering design tool. For the single-layer printed lines, the availability of approximate analytical closed-form expressions or diagrams for the dominant mode characteristics helps the designers of printed microwave circuits. Based on these, a microwave engineer may start the design of a printed structure. The structure may then be fine tuned through numerical electromagnetic simulation in conjunction with optimization techniques. In line with this approach, the characteristics of the dominant quasi-TEM mode will be presented. The approximate expressions for the effective dielectric constant (ε_{reff}) and the characteristic impedance (Z_0), by reviewing the works of Svacina [7], will be given.

Svacina, who adapted the Wheeler's classical works [8, 9], evaluated the effective permittivity (ε_{reff}) through a conformal transformation of a three-layer microstrip line. The remaining dimensional relations obtained through the conformal transformation, e.g., the characteristic impedance Z_0, the capacitance as well as the phase and group velocities, are not affected by the presence of the multiple layers [7]. Thus, their expressions (not their absolute values) remain the same as for the single-layer case, e.g., [3].

5.3 THREE–LAYERS MICROSTRIP LINE

Let us now elaborate on the three-layers microstrip line studied in [9]. The strip conductor may be located between the two dielectric layers (Figure 5.1a) or printed on the surface of the top dielectric layer (Figure 5.1b). These semi-infinite structures (z-plane) are first transformed to a finite (g-plane) plane according to [9], as shown in Figure 5.2. The strip and the ground-plane conductors are mapped to two-line segments, and the two-dielectric and air regions are mapped to corresponding areas ($S_{\varepsilon 1}$, $S_{\varepsilon 2}$, S_0) between the conductors. Each area is proportional to the electromagnetic energy confined within the corresponding region. The filling factors (q_1, q_2) of each dielectric layer are defined in the transformed complex plane as the ratio of the area assigned to each layer ($S_{\varepsilon 1}$, $S_{\varepsilon 2}$) to the total cross-section $S_c = S_0 + S_{\varepsilon 1} + S_{\varepsilon 2}$ between the strip conductors (Figure 5.2). Recall that the filling factor represents the degree by which electromagnetic power is confined within the corresponding layer [8, 9]. In this manner, the two filling factors take the following form [7]:

$$q_1 = \frac{S_{\varepsilon 1}}{S_c} = 1 - \frac{S_0 + S_{\varepsilon 2}}{S_c} \tag{5.1a}$$

$$q_2 = \frac{S_{\varepsilon 2}}{S_c} = 1 - \frac{S_0 + S_{\varepsilon 1}}{S_c} = 1 - q_1 - \frac{S_0}{S_c}. \tag{5.1b}$$

The filling factors are then given in [7] as follows:

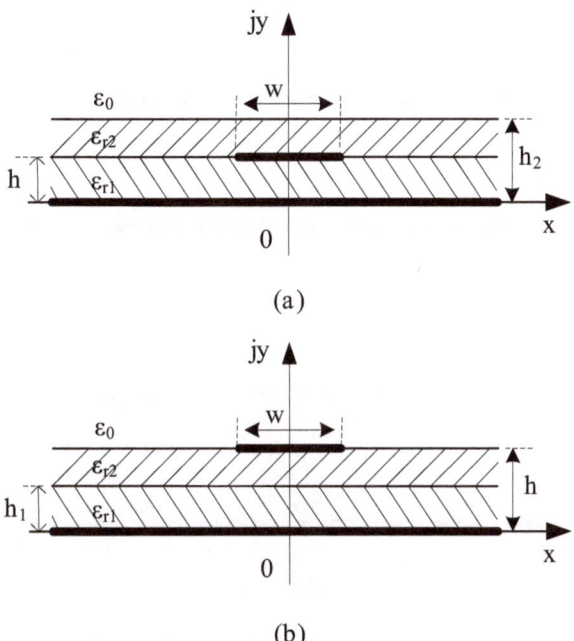

Figure 5.1: Three-dielectric-layers microstrip line with the strip conductor located: (a) between the two dielectrics and (b) printed on the top dielectric layer. This is defined as the original z-plane, $(z = x + jy)$.

Microstrip conductor at the dielectric interface (Figure 5.1a, Figure 5.2a)

Wide microstrips ($\bar{w} = w/h \geq 1$):

$$q_1 = 1 - \frac{0.5}{\bar{w}_{eff}} \cdot \ln\left(\pi \cdot \bar{w}_{eff} - 1\right) \tag{5.2a}$$

$$q_2 = 1 - q_1 - 0.5\frac{1 - \bar{U}_e}{\bar{w}_{eff}} \cdot \ln\left(\pi \cdot \bar{w}_{eff} \frac{\cos\left(0.5\pi\bar{U}_e\right)}{\pi\left(\bar{h}_2 - 0.5\right) + 0.5\pi\bar{U}_e} + \sin\left(0.5\pi\bar{U}_e\right)\right), \tag{5.2b}$$

where

$$\bar{w}_{eff} = \bar{w} + \frac{2}{\pi}\ln\left[17.08(0.5\bar{w} + 0.92)\right] \tag{5.2c}$$

$$\bar{U}_e = \frac{2}{\pi}\tan^{-1}\left[\frac{2\pi}{\pi\bar{w}_{eff} - 4} \cdot \left(\bar{h}_2 - 1\right)\right]. \tag{5.2d}$$

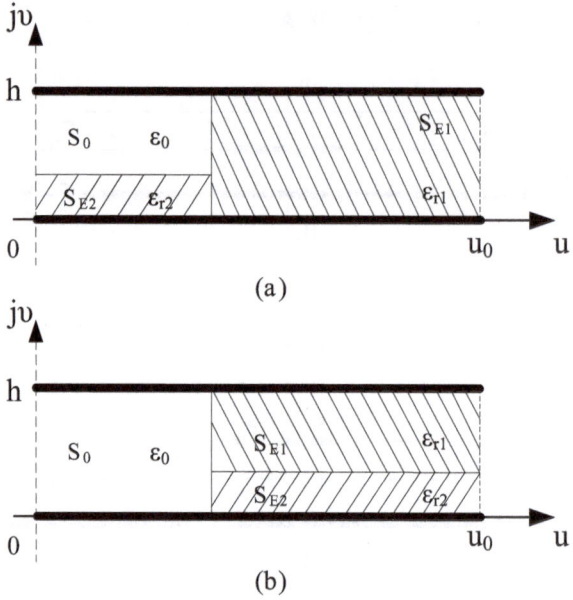

Figure 5.2: Conformal transformation of the three-layer microstrip line to the g-plane ($g = u + jv$) where the semi-infinite cross-section is mapped to a rectangular area between two conductor segments: (a) for Figure 5.1a and (b) for Figure 5.1b. These are approximated according to [7].

Narrow microstrips ($\bar{w} = w/h \leq 1$):

$$q_1 = 0.5 - \frac{0.9}{\pi \cdot \ln(\bar{w})} \tag{5.3a}$$

$$q_2 = 0.5 + \frac{1}{\pi \cdot \ln(8\bar{w})} \left\{ 0.9 + \frac{\pi}{4} \cdot \ln \left(\frac{\bar{h}_2 + 1}{\bar{h}_2 + 0.25\bar{w} - 1} \right) \cdot \right.$$

$$\left. \cos^{-1} \left[\left(1 - \frac{1}{\bar{h}_2} + 0.125 \frac{\bar{w}}{\bar{h}_2} \right) \sqrt{\frac{\bar{h}_2 + 1}{\bar{h}_2 + 0.25\bar{w} - 1}} \right] \right\}. \tag{5.3b}$$

The bar over each symbol represents values "normalized to h," e.g., $\bar{h}_2 = h_2/h$. With the availability of the filling factors q_1, q_2, the effective dielectric constant reads:

$$\varepsilon_{reff} = q_1\varepsilon_{r1} + \varepsilon_{r2} \frac{(1 - q_1)^2}{\varepsilon_{r2}(1 - q_1 - q_2) + q_2}. \tag{5.4}$$

Microstrip conductor printed on the top dielectric layer (Figure 5.1b, Figure 5.2b) [7]

Wide microstrips ($\bar{w} = w/h \geq 1$):

$$q_1 = 0.5\bar{h}_1 \left\{ 1 + \frac{\pi}{4} - \frac{1}{\bar{w}_{eff}} \ln \left[\pi \cdot \bar{w}_{eff} \frac{\sin(0.5\pi\bar{h}_1)}{0.5\pi\bar{h}_1} + \cos(0.5\pi\bar{h}_1) \right] \right\} \qquad (5.5a)$$

$$q_2 = 1 - q_1 - \frac{0.5}{\bar{w}_{eff}} \cdot \ln \left(\pi \cdot \bar{w}_{eff} - 1 \right), \qquad (5.5b)$$

where w_{eff} is given again by (5.2c).

Narrow microstrips ($\bar{w} = w/h \leq 1$):

$$q_1 = \ln \left(\frac{1 + \bar{h}_1}{1 - \bar{h}_1 + 0.25\bar{w}} \right)$$

$$\left\{ 1 + \frac{\pi}{4} - 0.5 \cos^{-1} \left[\left(0.125 \frac{\bar{w}}{\bar{h}_1} \right) \sqrt{\frac{1 + \bar{h}_1}{1 - \bar{h}_1 + 0.25\bar{w}}} \right] \right\} / \left[2 \ln \left(\frac{8}{\bar{w}} \right) \right] \qquad (5.6a)$$

$$q_2 = \frac{1}{2} - q_1 + \frac{0.9}{\pi \cdot \ln \left(\frac{8}{\bar{w}} \right)}. \qquad (5.6b)$$

The effective dielectric constant is:

$$\varepsilon_{reff} = 1 - q_1 - q_2 + \varepsilon_{r1}\varepsilon_{r2} \frac{(q_1 + q_2)^2}{\varepsilon_{r1}q_2 + \varepsilon_{r2}q_1}. \qquad (5.7)$$

The characteristic impedance expression remains the same as that for the single layer microstrip line, namely [3]:

$$Z_0 = \frac{Z_{0,air}}{\sqrt{\varepsilon_{reff}}}. \qquad (5.8a)$$

$$Z_0 = \frac{120\pi}{\sqrt{\varepsilon_{reff}}} \cdot \frac{1}{\bar{w}_{eff}} \qquad \text{for } \bar{w} = w/h \geq 1 \qquad (5.8b)$$

$$Z_0 = \frac{60}{\sqrt{\varepsilon_{reff}}} \cdot \ln \frac{8}{\bar{w}} \qquad \text{for } \bar{w} = w/h \leq 1. \qquad (5.8c)$$

Equation (5.8a) is explicitly given in Section 5.12.1, but based on that, Eqs. (5.8b) and (5.8c) are extracted. Svacina [7] tested his expressions against the numerical results [5, 6], as well as against published results for inverted microstrips [10, 11]. Always the differences observed were less than 2%.

5.4   MULTIPLE DIELECTRIC LAYER MICROSTRIP LINE

Svacina in [12] extended his conformal mapping method to multilayer microstrips and coplanar waveguides. The N-dielectric layer cross-section with the strip conductor printed on the top of the Mth layer is shown in Figure 5.3a. Figure 5.3b presents the conformal transformation of a multilayer strip.

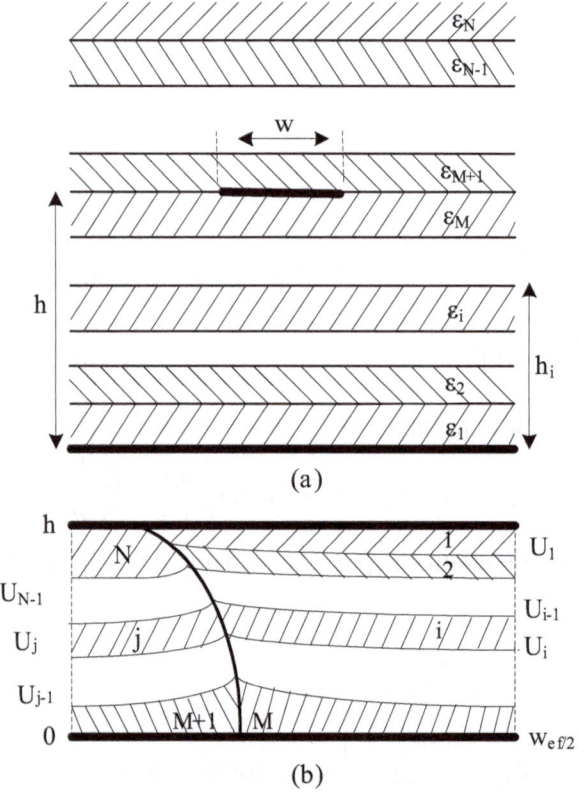

Figure 5.3: Multiple-dielectric-layer microstrip line: (a) the geometry and (b) the conformal transformation [12].

The filling factor expressions (q_i, q_j) are quite similar to Eqs. (5.2), (5.3), (5.5), and (5.6), where again the bar above each symbol represents normalization with respect to the strip conductor height (h) from the ground plane. The strip is assumed on the top of the Mth layer ($h = h_M$) and the quantities involved are:

$$\bar{w}_{eff} = \bar{w} + \frac{2}{\pi} \ln\left[17.08(0.5\bar{w} + 0.92)\right] \tag{5.9a}$$

$$\bar{U}_j = \frac{2}{\pi} \tan^{-1}\left[\frac{2\pi}{\pi \bar{w}_{eff} - 4} \cdot (\bar{h}_j - 1)\right] \tag{5.9b}$$

$$\bar{b}_j = \frac{\bar{h}_j + 1}{\bar{h}_j - 1 + 0.25\bar{w}} \tag{5.9c}$$

$$\bar{a}_i = \frac{\bar{h}_i + 1}{1 - \bar{h}_i + 0.25\bar{w}}. \tag{5.9d}$$

Filling factors for wide microstrips, $\bar{w} = w/h \geq 1$:

- Layers below the strip conductor $i = 1$ to M.

$$q_1 = q_k\left(\bar{h}_k = \bar{h}_1\right) \tag{5.10a}$$

$$q_i = q_k\left(\bar{h}_k = \bar{h}_i\right) - q_k\left(\bar{h}_k = \bar{h}_{i-1}\right) \quad \text{for } i = 2, 3, \ldots, M - 1, \tag{5.10b}$$

where

$$q_k\left(\bar{h}_k\right) = 0.5\bar{h}_k \left\{1 + \frac{\pi}{4} - \frac{1}{\bar{w}_{eff}} \ln\left[2\bar{w}_{eff}\frac{\sin\left(0.5\pi\bar{h}_k\right)}{\bar{h}_k} + \cos\left(0.5\pi\bar{h}_k\right)\right]\right\} \tag{5.10c}$$

$$q_M = 1 - \frac{1}{2\bar{w}_{eff}} \ln\left(\pi \cdot \bar{w}_{eff} - 1\right) - q_k\left(\bar{h}_k = \bar{h}_{M-1}\right). \tag{5.10d}$$

- Layers above the strip conductor $j = M + 1$ to N:

$$q_{M+1} = q_n\left(\bar{h}_n = \bar{h}_M, \bar{U}_h = \bar{U}_M\right) \tag{5.11a}$$

$$q_j = q_n\left(\bar{h}_n = \bar{h}_j, \bar{U}_n = \sqrt{\bar{U}_j}\right) - q_n\left(\bar{h}_n = \bar{h}_{j-1}, \bar{U}_n = \sqrt{\bar{U}_{j-1}}\right). \tag{5.11b}$$

For $j = M + 2, M + 3, \ldots$ to $N - 1$.

$$q_n \left(\bar{h}_n, \bar{U}_n \right) = \frac{1}{2\bar{w}_{\text{eff}}} \left\{ \ln \left(\pi \cdot \bar{w}_{\text{eff}} - 1 \right) - \left(1 - \bar{U}_n \right) \right.$$

$$\left. \ln \left[2\bar{w}_{\text{eff}} \frac{\cos \left(0.5\pi \bar{U}_n \right)}{2\bar{h}_n - 1 + \bar{U}_n} + \sin \left(0.5\pi \bar{U}_n \right) \right] \right\} \qquad (5.11c)$$

and

$$q_N = 1 - \sum_{v=1}^{N-1} q_v. \qquad (5.11d)$$

Filling factors of the narrow microstrips, $\bar{w} = w/h \leq 1$:

• Layers below the strip conductor $i = 1$ to M.

$$q_1 = q_k \left(\bar{h}_k = \bar{h}_1, \bar{\alpha}_k = \bar{\alpha}_1 \right) \qquad (5.12a)$$

$$q_i = q_k \left(\bar{h}_k = \bar{h}_i, \bar{\alpha}_k = \bar{\alpha}_i \right) - q_k \left(\bar{h}_k = \bar{h}_{i-1}, \bar{\alpha}_k = \bar{\alpha}_{i-1} \right)$$
$$\text{for } i = 2, 3, \ldots, M - 1, \qquad (5.12b)$$

where

$$q_k \left(\bar{h}_k, \bar{\alpha}_k \right) = -\frac{\ln \left(\bar{\alpha}_k \right)}{2 \ln \left(0.125\bar{w} \right)} \left\{ 1 + \frac{\pi}{4} - 0.5 \cos^{-1} \left[\frac{\bar{w}}{8\bar{h}_k} \sqrt{\bar{\alpha}_k} \right] \right\} \qquad (5.12c)$$

$$q_M = 0.5 - \frac{0.9}{\pi \cdot \ln \left(0.125\bar{w} \right)} - q_k \left(\bar{h}_k = \bar{h}_{M-1}, \bar{a}_k = \bar{a}_{M-1} \right). \qquad (5.12d)$$

• Layers above the strip conductor $j = M + 1$ to N:

$$q_{M+1} = q_n \left(\bar{h}_n = \bar{h}_{M+1}, \bar{b}_n = \bar{b}_{M+1} \right) \qquad (5.13a)$$

$$q_j = q_n \left(\bar{h}_n = \bar{h}_j, \bar{b}_n = \bar{b}_j \right) - q_n \left(\bar{h}_n = \bar{h}_{j-1}, \bar{b}_n = \bar{b}_{j-1} \right)$$
$$\text{for } j = M + 2, M + 3, \ldots, N - 1, \qquad (5.13b)$$

where

$$q_n \left(\bar{h}_b, \bar{b}_n \right) = 0.5 + \frac{1}{\pi \ln \left(0.125\bar{w} \right)}$$

$$\left\{ 0.9 + \frac{\pi}{4} \ln \left(\bar{b}_n \right) \cos^{-1} \left[\frac{\sqrt{\bar{b}_n}}{\bar{h}_n} \left(\bar{h}_n - 1 + 0.125\bar{w} \right) \right] \right\}. \qquad (5.13c)$$

The last layer q_N is again given by Eq. (5.11d).

Based on the transformed domain, (Figure 5.3b), Svacina extracted a general expression for the effective dielectric constant as:

$$\varepsilon_{reff} = \frac{\left(\sum_{i=1}^{M} q_i\right)^2}{\sum_{i=1}^{M} \left(\frac{q_i}{\varepsilon_{ri}}\right)^2} + \frac{\left(\sum_{i=M+1}^{N} q_j\right)^2}{\sum_{j=M+1}^{N} \left(\frac{q_j}{\varepsilon_{rj}}\right)^2}. \tag{5.14}$$

Numerous authors verified the above expression with accuracy better than 2%. In contrary Wan and Hoorfar [13] made a comparison against a moment method solution for a microstrip located on the first layer and for numbers of layers more than three. They found a large inaccuracy the order of 14–35% and suggested an improvement. However, a careful examination of [13] reveals that the authors misinterpreted the indeed difficult-to-read format of Svacina [12] expressions. Actually, the improvement suggested in [13] was already there, hidden in the [12] expressions.

Characteristic Impedance

A relatively simple formula represents Z_0 in terms of the effective dielectric constant as [3]:

$$Z_0(f \to 0) = \frac{Z_{0,air}}{\sqrt{\varepsilon_{reff}(f = 0)}}, \tag{5.15a}$$

where $Z_{0,air}$ is the characteristic impedance of the air-filled ($\varepsilon_r = \mu_r = 1$) microstrip:

$$Z_{0,air} = \begin{cases} 127.415 - 57.839 \ln{(\bar{w})} & \text{for } 0.05 \le \bar{w} \le 2 \\ 136.6131 - 30.88\bar{w} + 33.43\bar{w}^2 - 0.13384\bar{w}^3 & \text{for } 2 \le \bar{w} \le 10 \end{cases}. \tag{5.15b}$$

An alternative expression for $Z_{0,air} = Z_{LO}$ is given in [3] which is originally proposed by Hammerstad and Jensen [14].

5.5 FREQUENCY DISPERSION OF MULTILAYER MICROSTRIP LINES

The effective dielectric constant given in the previous sections is based on a conformal transformation which is exactly valid in the static case ($f = 0$ Hz), to be symbolized as $\varepsilon_{reff}(f = 0)$. Through a quasi-static approach, the above has an acceptable accuracy for low microwave frequencies ($f \to 0$). However, microstrip lines are frequency dispersive and their effective dielectric constant depends on frequency, $\varepsilon_{reff}(f)$. Verma and Sadr [15] proposed a dispersion model for multilayer structures by adapting the single-layer model of Kirschning and Jansen [16], which reads:

$$\varepsilon_{reff}(f) = \varepsilon_r - \frac{\varepsilon_r - \varepsilon_{reff}(f = 0)}{1 + P(f)}. \tag{5.16}$$

The accuracy of this model depends primarily on the accuracy of $\varepsilon_{reff}(f = 0)$. The term $P(f)$ accounts for the frequency dependence (given next in Eqs. (5.22)) and was estimated by curve fitting of measured or accurate numerical results from electromagnetic simulators, which accounts for the actual hybrid mode propagation. However, to apply (5.16), the multilayer structure must be reduced to an *equivalent Singler–Layer Reduction, SLR structure*. Note that the SLR approach enables the modeling of losses "dielectric loss" and "conductor loss" through the well-established closed form expressions for the single layer, e.g., Hoffmann [3].

5.6   EQUIVALENT SINGLE–LAYER MICROSTRIP (SLR)

The SLR initially proposed by Verma and Sadr [15] was further elaborated on by Verma and Bhupal [17, 18]. The multilayer microstrip line of Figure 5.4a is reduced, for example, to an equivalent single-layer microstrip shown in Figure 5.4b. The equivalent substrate thickness is given by the total height of the layers below the strip conductor (measured from the ground plane):

$$h_{eq} = h = h_M = \sum_{i=1}^{M} d_i, \tag{5.17}$$

while the equivalent strip width remains the same, $w_{eq} = w$.

The basic idea for the evaluation of an equivalent complex dielectric constant (ε_{req}^*) exploits an already available quasi-static effective dielectric constant (ε_{reff}). For this purpose, the expressions given in the previous sections, e.g., Eq. (5.16), can be used. Assuming that ε_{reff} is identical for both the original multilayer and the equivalent single-layer structures [19], the classical Wheeler's formulae apply for the latter case. Hence, a microstrip line with substrate thickness $h_{eq} = h$, complex dielectric constant (ε_{req}^*) and strip width $w_{eq} = w$ has the following complex effective dielectric constant [9]:

$$\varepsilon_{reff}^* = 1 + q\left(\varepsilon_{req}^* - 1\right), \tag{5.18}$$

where q is the filling factor of the multilayer structure that is assumed the same as that of the equivalent single layer. Thus, it can be evaluated according to [9] or any microstrip textbook, e.g., [3]. That is, the equivalent filling factor q is independent of the multi-layer dielectric constants and to a first approximation independent of ε_{req}. Since the already-given Svacina's anal-

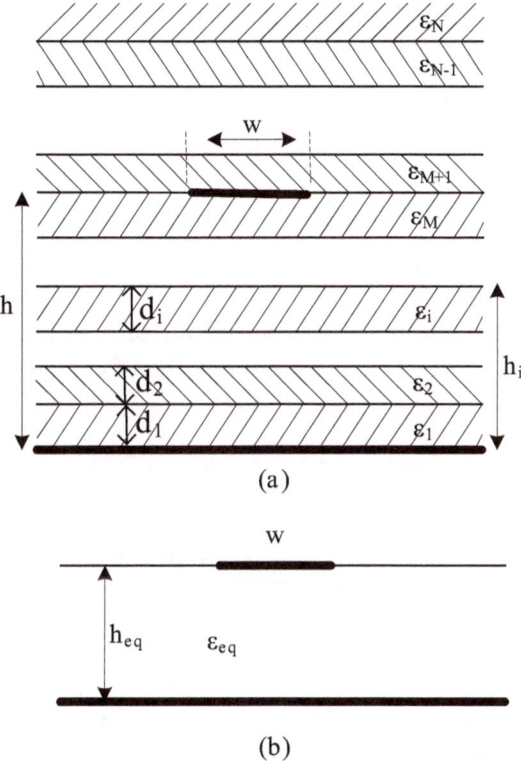

Figure 5.4: (a) A multilayer microstrip line and (b) reduction to an equivalent single layer.

ysis [7] is just an extension of that of Wheeler's [9], it is obvious that q can be obtained from Eqs. (5.2a) and (5.3a).

$$q = 1 - \frac{0.5}{\bar{w}_{eff}} \cdot \ln\left(\pi \cdot \bar{w}_{eff} - 1\right) \qquad \text{for } \bar{w} = w/h \geq 1 \qquad (5.19a)$$

$$q = 0.5 - \frac{0.9}{\pi \cdot \ln(\bar{w})} \qquad \text{for } \bar{w} = w/h \leq 1, \qquad (5.19b)$$

where $\bar{w}_{eff}$ is still given by (5.2c) and $w = w_{eq}, h = h_{eq}$.

Alternatively, the original Wheeler's formulae [9], repeated in [15], reads:

$$q = 0.5 + 0.5 \left(1 + \frac{12}{\bar{w}}\right)^{-1/2} \qquad \text{for } \bar{w} = w/h \geq 1 \qquad (5.20a)$$

$$q = 0.5 + 0.5 \left(1 + \frac{12}{\bar{w}}\right)^{-1/2} + 0.02(1 - \bar{w})^2 \qquad \text{for } \bar{w} = w/h \leq 1. \qquad (5.20b)$$

Instead of following [9], the effective (real part) dielectric constant given in Eq. (5.16) can estimate the equivalent (real part) dielectric constant. Therefore, rewriting (5.18), we have:

$$\varepsilon_{reff}(f = 0) = 1 + q\left(\varepsilon_{req} - 1\right) \qquad (5.21a)$$

or

$$\varepsilon_{req} = 1 + \frac{\varepsilon_{reff}(f = 0) - 1}{q}. \qquad (5.21b)$$

An approximation for the frequency dependence of the effective dielectric constant could be obtained with the aid of the Kirschning and Jansen [16] model presented in Eq. (5.16) by substituting $\varepsilon_r = \varepsilon_{req}$ through (5.21b), where $P(f)$ reads [3, 16]:

$$P(f) = P_1 P_2 \left[F \cdot H \cdot (0.1844 + P_3 P_4)\right]^{1.5763} \qquad (5.22a)$$

$$P_1 = 0.27488 + \left[0.6315 + \frac{0.525}{(1 + 0.157 \cdot F \cdot H)^{20}}\right] \cdot \bar{w} - 0.065683 \cdot e^{-8.7513\bar{w}} \qquad (5.22b)$$

$$P_2 = 0.033622 \left(1 - e^{-0.03442 \cdot \varepsilon_r}\right) \qquad (5.22c)$$

$$P_3 = 0.0363 e^{-4.6\bar{w}} \left\{1 - e^{\left(\frac{F \cdot H}{38.7}\right)^{4.97}}\right\} \qquad (5.22d)$$

$$P_4 = 1 + 2.751 \left\{1 - e^{-\left(\frac{\varepsilon_r}{15.916}\right)^8}\right\}. \qquad (5.22e)$$

The quantities are given as: $F = f$ GHz, $H = h$ mm, $\varepsilon_r = \varepsilon_{req}$, $\eta = \eta_\varepsilon$. Another possibility is to employ Kobayashi's expression [20]:

$$\varepsilon_{reff}(f) = \varepsilon_r - \frac{\varepsilon_r - \varepsilon_{reff}(f = 0)}{1 + (f/f_{50})^m} \qquad (5.23a)$$

where

$$f_{50} = \frac{C \cdot \tan^{-1}\left[\varepsilon_r \sqrt{\frac{\varepsilon_{reff}(f=0)-1}{\varepsilon_r - \varepsilon_{reff}(f=0)}}\right]}{2\pi h \sqrt{\varepsilon_r - \varepsilon_{reff}(f=0)} \cdot \left[0.75 + \bar{w}\left(0.75 - 0.332/\varepsilon_r^{1.73}\right)\right]} \tag{5.23b}$$

$$m = m_0 \cdot m_c \le 2.32 \tag{5.23c}$$

$$m_0 = 1 + \frac{1}{1+\sqrt{\bar{w}}} + 0.32\left(1+\sqrt{\bar{w}}\right)^{-3} \tag{5.23d}$$

$$m_c = \begin{cases} 1 + \frac{1.4}{1+\bar{w}}\left(0.15 - 0.235 \cdot e^{-0.45f/f_{50}}\right) & \text{for } \bar{w} \le 0.7 \\ 1 & \text{for } \bar{w} \ge 0.7. \end{cases} \tag{5.23e}$$

The above Kirschning and Jansen [16] and Kobayashi [20] expressions have an accuracy of better than 0.6% for the single-layer microstrip.

5.7 CHARACTERISTIC IMPEDANCE VS. FREQUENCY, $Z_0(f)$

The frequency dependence of the characteristic impedance can be evaluated using COMPEL's formulae by Tounsi et al. [21]:

$$Z_0(f) = Z_0(f \to 0)\frac{\varepsilon_{reff}(f)-1}{\varepsilon_{reff}(f=0)-1}\sqrt{\frac{\varepsilon_{reff}(f=0)}{\varepsilon_{reff}(f)}}, \tag{5.24}$$

where $Z_0(f \to 0)$ is the quasistatic characteristic impedance often found in [3] as a closed-form expression, or is given in diagrams and alternatively defined in Eq. (5.15a).

5.8 DIELECTRIC LOSSES

The evaluation of the attenuation factor due to dielectric losses can be based on the complex dielectric constant approach defined in (5.18). For this purpose, Verma et al. [15, 17, 18] adopt Yamashita's variational approach [5] in conjunction with the complex admittance function of the transverse transmission line [19]. However, this is a rather complicated technique and is restricted to a specific number of layers.

According to the classical microwave integrated-circuits handbook [3] and the original references therein, an equivalent effective dielectric loss factor ($\tan \delta_{\varepsilon,eff}$) can be defined for a transversely inhomogeneous quasi-TEM line. This is given from a static approximation under a low-loss condition, namely a small dielectric loss tangent $\tan \delta_\varepsilon \ll 1$, and it reads:

$$\tan \delta_{\varepsilon,eff} = q_{\tan \delta_\varepsilon} \cdot \tan \delta_\varepsilon, \tag{5.25}$$

where $q_{\tan \delta_\varepsilon}$ is the filling factor of the dielectric loss. This should not be confused with the filling factors $(q_\varepsilon \leftarrow q)$ adopted for the definition of the effective dielectric constant. A good approximation for $q_{\tan \delta_\varepsilon}$ is given in [3] for printed lines. In these lines, the dielectric boundary and the electric field are parallel in regions of high field, but not over the entire circuit, which includes microstrip lines, as:

$$q_{\tan \delta_\varepsilon} = \frac{\varepsilon_r}{\varepsilon_{reff}} q_\varepsilon. \tag{5.26a}$$

The filling factor $(q_\varepsilon = q)$ for a multilayer structure is related to the equivalent dielectric constant $(\varepsilon_{req} \rightarrow \varepsilon_r)$ and its corresponding effective value (ε_{reff}) through Eq. (5.18). Thus, substituting Eq.(5.18) in (5.26a) yields:

$$q_\varepsilon = q = \left(\varepsilon_{reff} - 1\right) / \left(\varepsilon_{reff} - 1\right) \tag{5.26b}$$

and

$$q_{\tan \delta_\varepsilon} = \frac{\varepsilon_r}{\varepsilon_{reff}} \cdot \frac{\varepsilon_{reff} - 1}{\varepsilon_r - 1} = \frac{1 - (1/\varepsilon_{reff})}{1 - (1/\varepsilon_r)}. \tag{5.26c}$$

Equations (5.25) and (5.26b) can be used through this chapter for the estimation of an effective dielectric loss tangent and, through that, the associated attenuation constant (α_d). Since a quasi TEM mode can be treated as an ordinary TEM mode propagating in a transmission line with a dielectric constant (ε_{reff}), the (α_d) can be readily adapted from this propagation mode, e.g., Pozar [1]:

$$\alpha_d = \frac{k \tan \delta_{\varepsilon,eff}}{2} = \frac{k_0 \sqrt{\varepsilon_{reff}} \tan \delta_{\varepsilon,eff}}{2}, \tag{5.27}$$

where k_0 is the free-space wavenumber.

5.9   COPLANAR TRANSMISSION LINES

The coplanar transmission lines are composed of strip conductors printed on the same plane, usually on the top surface of a dielectric substrate. Their established representatives are the coplanar waveguide (CPW), the Slotline, and the coplanar strip (CPS). All three of them offer a convenient shunt connection of microwave components, while CPW and CPS support both series and shunt connections.

Slotlines can be realized by etching a longitudinal slot (or the entire slotline circuit) on the metalized surface of a dielectric substrate, while its opposite side is left free (unmetallized). Its most interesting application refers to the hybrid microstrip-slotline circuits, where the slotline circuit is etched on the ground plane of the microstrip. In this manner, high characteristic impedance offered by the slotline can be exploited in the realization of high impedance series stubs and baluns (this is not possible in microstrip) as well as resonant and non-resonant antennas, or as the feed-excitation of printed antennas. Moreover, this hybridization may offer novel designs such as hybrid branchline couplers proposed by De Ronde [22]. In this configuration, the two series-connected microstrip branches are "eliminated" since they are realized by the slotline in the ground plane, leading to a substantial size reduction but also offering a wider bandwidth. More details can be found in [4] and the references therein.

The coplanar waveguide and coplanar strips geometries shown in Figures 5.5b and c can be understood as being complementary to each other.

Considering the strip conductors as PEC (perfect electric conductor) films and the free substrate surface being approximated by PMC (perfect magnetic conductor) films, the slotline is identically complementary to the conventional microstrip line. In turn, considering only the metallization plane, CPW and CPS can be characterized as complementary. This approach enables the employment of the same conformal transformation for their analysis, as will be described later on.

The CPW was proposed by Wen [23] as early as 1969. However, the convenience offered in the fabrication of MMICs was established later on. The specific advantage of CPW stems from the presence of the signal and ground conductors on the same plane offering series and shunt connections. These connections are made without the need of via holes through the metallization to reach the ground plane on the other side of the substrate, as required by the established microstrip technology. Hence, CPW technology could eliminate the complicated "back-plane mask processing" required in microstrip technology, and its substitution was constantly sought by MMIC foundries. Actually, a combination of these technologies is used since microstrip technology is very well studied and almost standardized.

CPS technology shares the series and shunt connection features common with CPW. However, using CPS, it is easier to realize higher characteristic impedances than with CPW regardless of the substrate thickness, sharing this feature with slotlines.

Moreover, symmetric (identical strips of the same width) CPS is an inherently balanced line. That happens because currents on the two strips are opposite but equal and have the same current density distribution. Their balanced nature is exploited in the design-realization of balanced circuits such as mixers and modulators. Moreover, their balanced nature makes them less sensitive to induced noises and thus appropriate for the low-noise amplifiers and low-phase noise oscillators. Overall, CPS's are widely used in high-speed digital and analog microwave and millimeter wave-integrated circuits; see [24] and the references therein. Fabrication tolerances may result in different strip widths yielding an "asymmetric CPS," which may degrade its balanced nature and its immunity to noise performance. However, asymmetric CPS may be desirable in some applications since its characteristic impedance can be controlled by an appropriate adjustment of the one strip width.

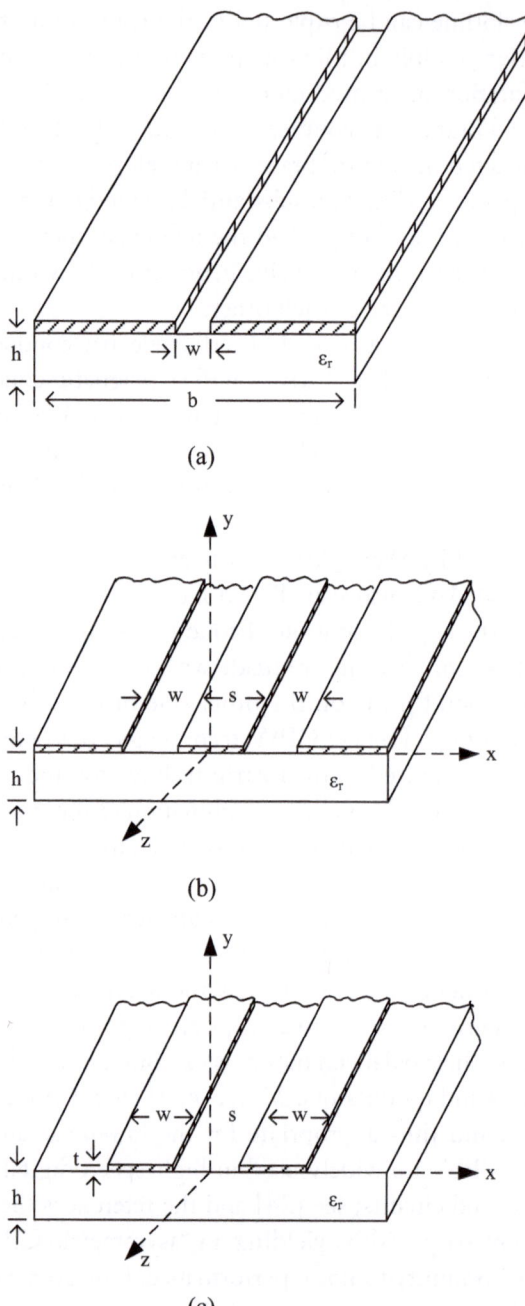

Figure 5.5: Geometry of coplanar transmission lines, (a) slotline, (b) coplanar waveguide, and (c) coplanar strips.

5.10 MULTILAYER COPLANAR WAVEGUIDE (CPW)

CPW technology is, in general, advantageous over microstrip in the convenience offered in the shunt connection of microwave components, in addition to the series connection supported by both. This is particularly important for microwave-integrated components. That is because it avoids the necessity of backplane processing, which is inevitable in microstrip technology shunt connections realized through electroplated ground plane via holes. However, CPW is inferior against microstrip in efficient interconnections and components modeling due to the involved complexity in its analytical electromagnetic analysis.

5.10.1 QUASI–STATIC APPROXIMATION

An asymmetric coplanar line printed in a multilayer structure is presented in Figure 5.6a. The signal strip width is s and is separated by different gaps w_1, w_2 from the left and the right ground-strip conductors. These strips are located between $M + 1$ dielectric layers from below and $N + 1$ layers from above. Svacina [12] performed a conformal Christoffel–Schwartz transformation, mapping this structure to the approximate equivalent formation of Figure 5.6b. The coordinates in the transformed w-plane can be expressed in terms of the complete elliptic integrals of the first kind $K(k)$ and $K'(k) = K(k')$ as [25]:

$$w = u + jv = K(k) + jK'(k) = K(k) + jK(k'), \qquad (5.28a)$$

where $k' = \sqrt{1 - k^2}$ and the properties of elliptic integrals are elaborated in detail in standard mathematical handbooks, e.g., [25].

Concerning the two ground strips of Figure 5.6a, these are mapped on the segments at $v = \pm v_0$, while the signal conductor is mapped at $v = 0$. Likewise, the layers $i = 1$ to M and $j = 1$ to N are mapped to a pair of symmetric strips, while the semi-infinite (usually air $\varepsilon_r = \mu_r = 1$) layers $i = M + 1$ and $j = N + 1$ are mapped to two centrally located areas. The dimensions of each ith-mapped layer is approximated as $2v_0(u_1 - u_{i-1})$, where these coordinates are defined by Eq. (5.28a) and their ratios are:

$$\frac{v_0}{u_0} = 2\frac{K(k')}{K(k)}, \qquad \frac{v_0}{u_i} = 2\frac{K(k_i')}{K(k_i)} \qquad (5.29a)$$

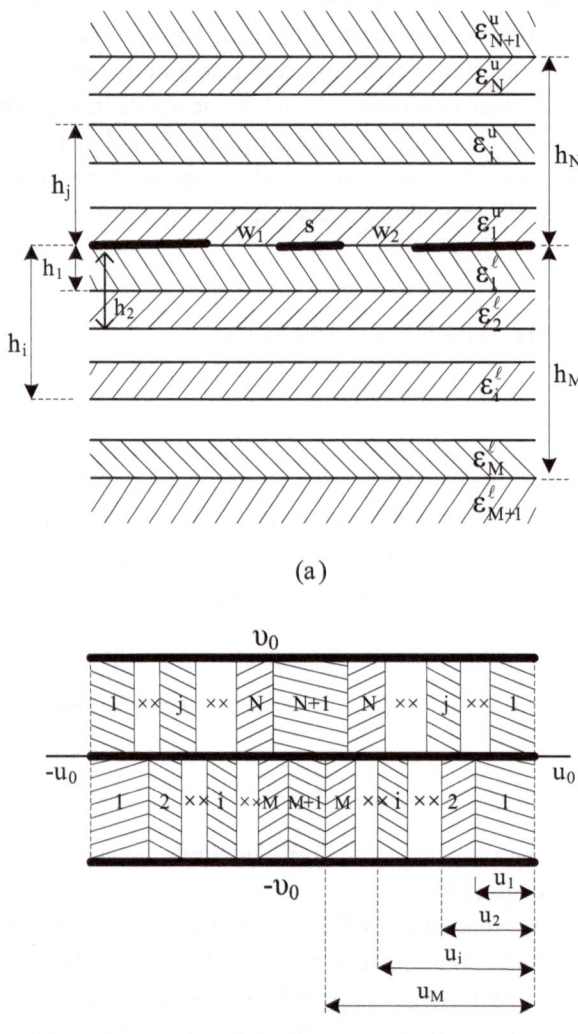

(a)

(b)

Figure 5.6: A multilayer asymmetric coplanar waveguide: (a) original $z = x + jy$ plane and (b) approximate conformal transform in $w = u + jv$ domain.

where

$$k_i^2 = 2\frac{k_{i1} + k_{i2}}{(1 + k_{i1})(1 + k_{i2})} \tag{5.29b}$$

$$k_{i1} = \frac{\sinh\left(\frac{\pi}{4} \cdot \frac{s}{h_i}\right)}{\sinh\left(\frac{\pi}{4h_i}(s + 2w_1)\right)} \tag{5.29c}$$

$$k_{i2} = \frac{\sinh\left(\frac{\pi}{4} \cdot \frac{s}{h_i}\right)}{\sinh\left(\frac{\pi}{4h_i}(s + 2w_2)\right)} \tag{5.29d}$$

and $k_0 = 0$, while for the semi-infinite top and bottom layers:

$$k = k_{M+1} = k_{N+1} = \lim_{h_i = h_{M+1} = h_{N+1} \to 0} k_i = \frac{s(s + w_1 + w_2)}{(s + w_1)(s + w_2)}. \tag{5.29e}$$

The latter is obtained by applying De L'Hospital's rule on Eqs.(5.29c) and (5.29d) to yield

$$k_{01} = \frac{s}{s + 2w_1} \quad \text{and} \quad k_{02} = \frac{s}{s + 2w_2}. \tag{5.29f}$$

In view of the above, the filing factor (q_i) for each ith layer is defined by the ratio of its assigned area to the total cross-section $(2u_0 \cdot 2v_0 = 4u_0v_0)$, which using (5.29a) reads:

$$q_i = \frac{2v_0(u_i - u_{i-1})}{4u_0 \cdot v_0} = \frac{1}{2}\frac{v_0}{u_0}\left[\frac{u_i}{v_0} - \frac{u_{i-1}}{v_0}\right] \quad \text{or} \tag{5.30a}$$

$$q_i = \frac{1}{2}\frac{K(k')}{K(k)} \cdot \left[\frac{K(k_i)}{K(k_i')} - \frac{K(k_{i-1})}{K(k_{i-1}')}\right]. \tag{5.30b}$$

Equation (5.30b) applies for both $i = 1$ to $M + 1$ and $i \to j = 1$ to $N + 1$ layers and $k_0 = 0$.

The effective dielectric constant (quasi-static expression) results from a summation of the product $(q_i \varepsilon_{ri})$ over all layers as:

$$\varepsilon_{reff} = \sum_{i=1}^{M+1} q_i e_{ri}^l + \sum_{j=1}^{N+1} q_j e_{rj}^u \tag{5.31}$$

$l = $ lower, $u = $ upper.

The above expressions can be significantly simplified in the symmetric CPW case, which is worth a separate section.

5.11   SYMMETRIC MULTILAYER COPLANAR WAVEGUIDES $(w_1 = w_2 = w)$

The expressions of the previous section obviously apply to the symmetric CPW by enforcing the gaps between the signal conductor and the ground conductors to become equal as $w_1 = w_2 = w$. However, they can be greatly simplified by exploiting the complete elliptical integrals K and K' properties. For this purpose, let us start from Eqs. (5.29c) and (5.29d) to write:

$$k_{i1} = k_{i2} = k_{i\varepsilon} = \frac{\sinh\left(\frac{\pi}{4} \cdot \frac{s}{h_i}\right)}{\sinh\left(\frac{\pi}{4h_i}(s + 2w)\right)}. \tag{5.32a}$$

In turn k_i and k_i' read:

$$k_i = \frac{2\sqrt{k_{i\varepsilon}}}{1 + k_{i\varepsilon}} \tag{5.32b}$$

$$k_i' = \sqrt{1 - k_i^2} = \frac{1 - k_{i\varepsilon}}{1 + k_{i\varepsilon}}. \tag{5.32c}$$

For the two semi-infinite top and bottom layers $h_{M+1}, h_{N+1} \to \infty$, let us take the limit of (5.32a) by applying De L'Hospital's rule to yield:

$$k_\varepsilon = \lim_{h_i \to \infty} k_{i\varepsilon} = \frac{s}{s + 2w} \quad \text{and} \tag{5.32d}$$

$$k = \frac{2\sqrt{k_\varepsilon}}{1 + k_\varepsilon}. \tag{5.32e}$$

In view of Eqs. (5.32), the expressions (5.30) for the filling factors can be simplified with the aid of the K and K' properties given in the Table of Integrals of [26] as:

$$K\left(\frac{2\sqrt{k}}{1 + k}\right) = (1 + k)K(k) \tag{5.33a}$$

$$K\left(\frac{1 - k'}{1 + k'}\right) = \frac{1 + k'}{2}K(k) \quad \text{where } k' = \sqrt{1 - k^2} \text{ or} \tag{5.33b}$$

$$K\left(\frac{1 - k}{1 + k}\right) = \frac{1 + k}{2}K(k') \quad \text{where } K(k') = K'(k). \tag{5.33c}$$

Equations (5.32) and (5.33) can be exploited for each of the three terms of the filling factor (5.30), as, for example:

$$\frac{K(k_i)}{K(k_i')} = \frac{(1 + k_{i\varepsilon})K(k_{i\varepsilon})}{(1/2)(1 + k_{i\varepsilon})K(k_{i-1,\varepsilon}')} = 2\frac{K(k_{i\varepsilon})}{K(k_{i-1,\varepsilon}')}. \tag{5.34}$$

Substituting in the same way, the three terms (5.30b) read:

$$q_i = \frac{1}{2} \cdot \frac{1}{2}\frac{K(k_\varepsilon')}{K(k_\varepsilon)} \left\{ 2\frac{K(k_{i\varepsilon})}{K(k_{i\varepsilon}')} - 2\frac{K(k_{i-1,\varepsilon})}{K(k_{i-1,\varepsilon}')} \right\}. \tag{5.35}$$

It is interesting to note that for the first layer $i = 1$, either above or below the strip conductors, the last term of Eqs. (5.30b) or (5.35) involves some limiting value problems since k_{i-1} or $k_{i-1,\varepsilon}$ becomes $k_0 = 0$ and $k_0' = 1$ [12]. The corresponding limiting values read [26]:

$$\lim_{m \to 0} K(m) = \frac{\pi}{2} \quad \text{and} \quad \lim_{m \to 1} K(m) = \infty. \tag{5.36a}$$

Hence,

$$\lim_{k_0 \to 0} \left\{ \frac{K(k_0)}{K(k_0')} \right\} = \frac{\frac{\pi}{2}}{\infty} = 0. \tag{5.36b}$$

Thus, the first filling factor reads

$$q_1 = \frac{1}{2}\frac{K(k_\varepsilon')}{K(k_\varepsilon)} \cdot \frac{K(k_{i\varepsilon})}{K(k_{i\varepsilon}')}. \tag{5.37}$$

Concerning the filling factors of the two semi-infinite layers $i = N + 1$ and $M + 1$, Eq. (5.32e) applies as: $k_{i\varepsilon} = k_\varepsilon$. Thus, Eq. (5.35) is simplified to:

$$q_i = \frac{1}{2} - \frac{1}{2}\frac{K(k_\varepsilon')}{K(k_\varepsilon)} \cdot \frac{K(k_{i-1,\varepsilon})}{K\left(k_{i-1,\varepsilon}'\right)}. \tag{5.38}$$

At this point it is very useful to tabulate some accurate enough and very simple asymptotic approximations for the above ratios of complete elliptic integrals according to [4, 24], and [28]:

$$\frac{K(k')}{K(k)} = \begin{cases} \frac{2}{\pi}\ln(4/k) & \text{for } 0 < k < 10^{-5} \\ \frac{1}{\pi}\ln\left[4\frac{1+k'}{1-k'}\right] = \frac{2}{\pi}\ln\left[2\frac{\sqrt{1+k'}}{\sqrt{1-k'}}\right] & \text{for } 10^{-5} < k \le \frac{1}{\sqrt{2}} \end{cases} \tag{5.39a}$$

$$\frac{K(k)}{K(k')} = \frac{1}{\pi}\ln\left[4\frac{1+k}{1-k}\right] = \frac{2}{\pi}\ln\left[2\sqrt{\frac{1+k}{1-k}}\right] \quad \text{for } \frac{1}{\sqrt{2}} \le k \le 1. \tag{5.39b}$$

It is now interesting to give the above expressions for the single-layer CPW.

5.11.1   SINGLE–LAYER CPW

The effective permittivity of an ordinary single-layer CPW can be obtained from Eqs. (5.30) and (5.31) by just allowing only one substrate ($M = 1$) and no superstrate layers ($N = 0$). Also, the infinitely extending layer above and below the structure must have air characteristics ($\varepsilon_{r2}^l = \varepsilon_{r1}^u = 1$). Moreover, for a symmetrical single-layer CPW, the result becomes identical to that of Ghione and Naldi [29]. For the effective dielectric constant, (5.31) reads:

$$\varepsilon_{reff} = q_{i1}\varepsilon_{r1}^l + q_{i2}\varepsilon_{r2}^l + q_{j1}\varepsilon_{r1}^u = q_{i2} + q_{j1} + q_{i1}\varepsilon_r. \tag{5.40}$$

For the first bottom layer $i = 1$, q_{i1} is given by (5.37), with k_ε defined in (5.32e), and $k_{i\varepsilon}$ by (5.32a) with $h_i = h_1 = h$. For the upper semi-infinite layer where $j = 1$, q_{j1} is defined by (5.38). The second term is zero according to (5.36b), since $k_{j-1} = 0$, thus, $q_{j1} = 1/2$. Likewise, for the bottom semi-infinite layer $i = 2$, q_{i2} is given by (5.38) with $i - 1 = 1$ or $k_{i-1,\varepsilon} = k_{i\varepsilon}$, which yields:

$$q_{i2} = \frac{1}{2} - q_{i1}. \tag{5.41}$$

In turn,

$$\varepsilon_{reff} = \frac{1}{2} + \left(\frac{1}{2} - q_{i1}\right) + q_{i1} \cdot \varepsilon_r = 1 + (\varepsilon_r - 1)\,q_{i1} \quad \text{or} \tag{5.42a}$$

$$\varepsilon_{reff} = 1 + \frac{\varepsilon_r - 1}{2} \cdot \frac{K\left(k_\varepsilon'\right)}{K\left(k_\varepsilon\right)} \cdot \frac{K\left(k_{i\varepsilon}\right)}{K\left(k_{i\varepsilon}'\right)}. \tag{5.42b}$$

The above result is identical to that of [29].

5.12   MULTILAYER CPW WITH FINITE GROUND PLANES

The characteristics of CPW with infinite ground planes are presented in the previous section, and the analysis given is mainly based on the work of Svacina [12]. The CPW with finite ground plane (Figure 5.7a), as well as the CPS transmission line (Figure 5.7b) is analyzed by Chen and Chou [30]. This work is also based on conformal transformation for its quasi-static analysis, but a slightly different approach is followed. That is, [30] extends the "superposition of partial capacitances" approximation proposed by Veyres and Fouad Hanna [31]. A careful comparison of the final expression of [29] for the effective dielectric constant against that of the infinite ground plane given in Eqs.(5.30) and (5.31) reveals that they are identical. Actually, [30] gives an expression for five layers. Hence, this can be extended to $(N + 1)$ plus $(M + 1)$ layers with a careful construction of an expression like Eqs.(5.30) and (5.31). The information regarding the finite ground planes is included in k_i expressions corresponding to those of Eqs (5.29) or (5.32) for the symmetric geometry.

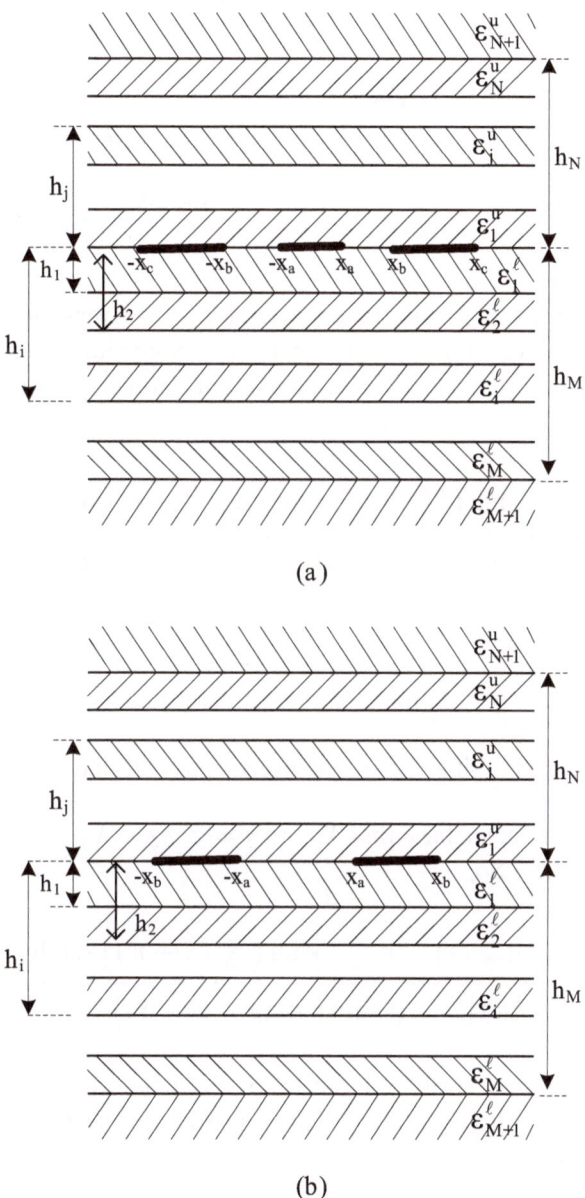

(a)

(b)

Figure 5.7: Generalized multilayer transmission lines with finite ground planes. (a) Coplanar waveguide (CPW) with finite ground planes and (b) coplanar strips (CPS).

Once again, we observe in Eqs. (5.35), (5.37), and (5.38) that the filling factors of successive layers $(i, i + 1)$ have a common term. Thus, the effective permittivity of (5.31) can be formulated as:

$$\varepsilon_{reff} = 1 + \frac{1}{2} \sum_{i=1}^{M} \left(\varepsilon_{ri}^{l} - \varepsilon_{ri+1}^{l} \right) \cdot \frac{K\left(k_{\varepsilon}'\right)}{K\left(k_{\varepsilon}\right)} \frac{K\left(k_{i\varepsilon}\right)}{K\left(k_{i\varepsilon}'\right)}$$

$$+ \frac{1}{2} \sum_{j=1}^{N} \left(\varepsilon_{rj}^{u} - \varepsilon_{rj+1}^{u} \right) \cdot \frac{K(k_{\varepsilon}')}{K(k_{\varepsilon})} \frac{K(k_{j\varepsilon})}{K(k_{j\varepsilon}')}. \tag{5.43}$$

Assuming two top $(M = 2)$ and three bottom $(N = 3)$ layers, we can see that Eq. (5.43) becomes identical to that given in [30]. Hence, the same expression applies for the finite ground plane CPW, but with the arguments of complete elliptic integrals substituted by those given in [30]:

$$k_i = \frac{\sinh\left(\frac{\pi X_c}{2h_i}\right)}{\sinh\left(\frac{\pi X_b}{2h_i}\right)} \cdot \sqrt{\frac{\sinh^2\left(\frac{\pi X_b}{2h_i}\right) - \sinh^2\left(\frac{\pi X_a}{2h_i}\right)}{\sinh^2\left(\frac{\pi X_b}{2h_i}\right) - \sinh^2\left(\frac{\pi X_a}{2h_i}\right)}}$$

$$k_i' = \sqrt{1 - k_i^2}. \tag{5.44}$$

Observing the geometry of Figure 5.7a, we can use the identity [25]:

$$\sinh^2(Z_1) - \sinh^2(Z_2) = \sinh(Z_1 + Z_2)\sinh(Z_1 - Z_2).$$

Also, from Figure 5.7a, we have that $X_c - X_b = g, X_b - X_a = w, X_b = w + S/2, X_c = g + w + S/2$ and $S = 2X_a$. So,

$$k_i = \frac{\sinh\left(\frac{\pi}{4h_i}(S + 2g + 2w)\right)}{\sinh\left(\frac{\pi}{4h_i}(S + 2w)\right)} \cdot \sqrt{\frac{\sinh\left(\frac{\pi}{2h_i}(S + w)\right) \cdot \sinh\left(\frac{\pi}{4h_i}w\right)}{\sinh\left(\frac{\pi}{2h_i}(g + w + S)\right) \cdot \sinh\left(\frac{\pi}{2h_i}(g + w)\right)}}. \tag{5.45}$$

Equation (5.45) can be exactly reduced to (5.32a) when the width of the ground plane tends to infinity $(g \to \infty)$. Also, for a single-layer substrate $(M = 1, N = 0)$, the effective permittivity expression (5.43) is reduced to Eq. (5.42b).

5.12.1   CHARACTERISTIC IMPEDANCE

According to the classical quasi-static approximation, e.g., Pozar [1] or Gupta et al. [4], the effective dielectric constant (ε_{reff}), the phase velocity (V_p), and the characteristic impedance (Z_0) can be written in terms of the line capacitance with and without the dielectric layers loading.

Explicitly:

$$\varepsilon_{reff} = \frac{C}{C_0} = \frac{\text{Capacitance per unit length with dielectric layers}}{\text{Capacitance per unit length with the strip conductors in the air}} \qquad (5.46)$$

$$V_p = \frac{C_{light}}{\sqrt{\varepsilon_{reff}}}, \qquad (5.47)$$

where C_{light} is the wave velocity (or light velocity) in free space ($C_{light} = 1/\sqrt{\mu_0 \varepsilon_0}$).

$$Z_0 = \frac{1}{C \cdot V_p} = \frac{\sqrt{\varepsilon_{reff}}}{C \cdot C_{light}} = \frac{1}{C \cdot C_0 \cdot \sqrt{\varepsilon_{reff}}}. \qquad (5.48)$$

From Eq. (5.48), one may observe that the only missing quantity is the quasi-static capacitance (C_0) per unit length when the strip conductors are immersed in air. Veyres and Fouad Hanna [32], using conformal mapping, evaluated C_0 as:

$$C_0 = 4\varepsilon_0 \frac{K'(k_0)}{K(k_0)} = 4\varepsilon_0 \frac{K(k_0')}{K(k_0)}. \qquad (5.49)$$

For a symmetric CPW with finite ground conductor as in Figure 5.6, K is defined in [32] and repeated in [30] as:

$$k_0 = \frac{X_c}{X_b} \sqrt{\frac{X_b^2 - X_a^2}{X_c^2 - X_a^2}} = \left(1 + \frac{2g}{S + 2w}\right) \cdot \left(\frac{w \cdot (s + w)}{(g + w) \cdot (g + s + w)}\right)^{1/2}. \qquad (5.50a)$$

For a symmetric CPW with infinite ground, k_0 can be obtained by taking the limit $g \to \infty$ in Eqs. (5.50), which is also given by [4] as:

$$k_0 = \frac{s}{s + 2w}. \qquad (5.50b)$$

Fouad Hanna and Thebault [32] gave the corresponding analysis for the asymmetric CPW and infinite ground strips:

$$k_0 = \frac{\frac{s}{2}\left[1 + \alpha\left(w_1 + \frac{s}{2}\right)\right]}{\frac{s}{2} + w_1 + \alpha \cdot \left(\frac{s}{2}\right)^2} \qquad (5.50c)$$

$$\alpha = \frac{w_1 w_2 + \frac{s}{2}(w_1 + w_2) \pm \{w_1 w_2(s + w_1)(s + w_2)\}^{1/2}}{\left(\frac{s}{2}\right)^2 (w_1 - w_2)}. \qquad (5.50d)$$

Considering the above definition of k_0, the characteristic impedance in every case can be obtained by substituting Eq. (5.49) into (5.48) and using the corresponding effective permittivity definitions from the previous sections as:

$$Z_0 = \frac{1}{C_{light} \cdot 4\varepsilon_0 \cdot \sqrt{\varepsilon_{reff}}} \cdot \frac{K(k_0)}{K(k_0')} = \frac{n_0}{4\sqrt{\varepsilon_{reff}}} \cdot \frac{K(k_0)}{K(k_0')} = \frac{30\pi}{\sqrt{\varepsilon_{reff}}} \cdot \frac{K(k_0)}{K(k_0')}, \quad (5.51)$$

where $C_{light} = 1/\sqrt{\mu_0 \varepsilon_0}$, $C_{light} \cdot \varepsilon_0 = 1/\sqrt{\varepsilon_0/\mu_0} = 1/\eta_0$ and $\eta_0 = 120\,\pi$ Ohms, the intrinsic (wave) impedance of free space (air).

5.13   MULTILAYER COPLANAR STRIPS

By carefully observing Figure 5.5, one may realize that CPS is complementary to the CPW with infinite ground strips at the metallization plane. That is, by assuming the two slots of width w to be replaced by strips and the strip conductors (signal strip of width s and the two infinite ground strips) to be removed, then the resulting geometry becomes identical to the CPS of Figure 5.6c. Thus, considering the thin strips as perfect electric walls (PEW) and the slots at the substrate surface being approximated by perfect magnetic walls (PMW), then the roles of PEW and PMW are interchanged between CPW and CPS as far as the substrate upper interface is concerned. However, this latter consideration (applying duality only for the strip conductor's surface) is enough for the application of the "duality" or "Babinet's" principle, e.g., Collin [33]. This is a direct consequence of the symmetry in Maxwell curl equations and is expressed as:

$$E^{CPS} = \sqrt{\mu/\varepsilon} \cdot H^{CPW} \qquad H^{CPS} = -\sqrt{\varepsilon/\mu} \cdot E^{CPW}. \quad (5.52)$$

It is obvious from Eq. (5.52) that the roles of electric and magnetic fields are interchanged, hence the corresponding boundary conditions should be interchanged as well. This requirement is already ensured at the top substrate surface by the interchange of electric and magnetic walls. What about the boundary conditions at the interface between different layers and the bottom layer with the air? Fortunately, at these interfaces, only the *continuity of both tangential electric and magnetic field is required*. Hence, these homogeneous boundary conditions remain the same between CPS and CPW. Based on the same reasoning, multiple layers can be added on top of the strip to form structures like that of Figure 5.6. In this case, the new CPW and CPS structures remain complementary (provided that the ground strips of CPW are semi-infinite). Concluding the above, the fields obtained by applying the conformal transformation in CPW can be readily exploited to obtain the CPS fields.

5.14   MICROSTRIP LINE ON A SINGLE MAGNETIC SUBSTRATE

The analysis of microstrip lines printed on a magnetized ferrite substrate constitute a rather complicated task due to the involved tensor magnetic permeability. Hence, the first pioneer works of Pucel and Masse [34, 35] for microstrip lines on magnetic substrates were restricted to the isotropic case with relative magnetic permeability μ_r and dielectric constant ε_r, as shown in Figure 5.8.

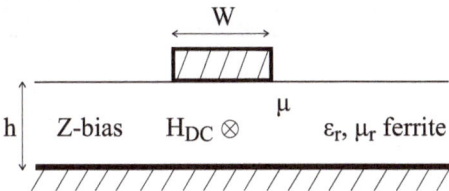

Figure 5.8: Microstrip line printed on a demagnetized or partially magnetized isotropic magnetic substrate [34].

Even though this approach seems useless for ferrite substrates, it is indeed appropriate for two practical cases when: (i) the substrate is demagnetized and (ii) the substrate is partially magnetized [35]. For the latter, the application concerns the ferrite "latched" state, that is, when the DC biasing corresponds to remnant magnetization (M_r is the magnetization retained after decreasing—removal of the DC biasing magnetic field to zero).

Just as in the dielectric substrate case, the dominant microstrip line mode can be considered as quasi-TEM. Thus, a transformation to a TEM line embedded in an equivalent effective medium characterized by (ε_{reff}, μ_{reff}) is required. While ε_{reff} is already available through the classical Wheeler's approach [9], the μ_{reff} should be estimated. For this purpose, the authors in [35] adopted a type of "duality principle" to exploit the analysis available for the dielectric substrate. It should be emphasized here that this is not the classical "duality principle" but a modified one applied only to TEM waves. For clarity reasons, both duality principles will be recalled herein. The classical one, e.g., Harrington [36], or Balanis [37], established a duality between "electric" and "magnetic" sources as given in the left part of Table 5.1. However, this correspondence between electric and magnetic quantities yields a mapping of the microstrip line to a slot line, as shown in Figure 5.9.

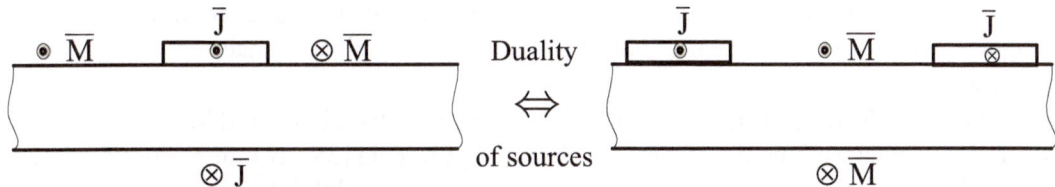

Figure 5.9: Transforming a microstrip line to a slotline based on the classical "duality principle."

Table 5.1: Duality principle: classical vs. TEM mode case

Classical Duality		TEM-mode Duality	
Electric Sources	Magnetic Sources	Electric Field	Magnetic Field
E_A	H_F	$\overline{E} = -\nabla\phi$	$\overline{B} = \nabla \times \overline{A} = -\hat{z} \times \nabla\psi_m$
H_A	$-E_F$	$\nabla^2\phi = -\rho/\varepsilon$	$\nabla^2\overline{A} = -\mu\overline{J} \to \nabla^2\psi_m = -\mu\overline{J}_z$
A	F	$\overline{J} = \hat{z}\overline{J}_z \to -j\omega\rho = \dfrac{\partial\overline{J}_z}{\partial z}$	$\overline{J} = \hat{z}\overline{J}_z \to \overline{A} = \hat{z}\overline{A}_z \ \& \ \psi_m = A_z$
ε	μ	ε_r	$1/\mu_r$
μ	ε	$1/\varepsilon_r$	μ_r
$k = \omega\sqrt{\mu\varepsilon}$	$k = \omega\sqrt{\mu\varepsilon}$	$\varepsilon_{reff} = \dfrac{C_{diel}}{C_{air}}$	$\dfrac{1}{\mu_{reff}} = \dfrac{L_{air}}{L_{magnetic}}$
$\eta = \sqrt{\mu/\varepsilon}$	$1/\eta$	$\phi_1 = \phi_2 \ \& \ \dfrac{\partial\phi_1}{\partial n} = \dfrac{\partial\phi_2}{\partial n}$	$\overline{A}_{z1} = \overline{A}_{z2} \ \& \ \dfrac{\partial\overline{A}_{z1}}{\partial n} = \dfrac{\partial\overline{A}_{z2}}{\partial n}$
$1/\eta$	η	C	$1/L$
$\upsilon_p = 1/\sqrt{\mu\varepsilon}$	υ_p	$V = Q/C$	$\Phi = LI = \int_s \overline{B} \cdot d\overline{S} = \oint_c \overline{A} \cdot d\overline{l}$
J	M	$Q = \int_s \rho\,dy'$	$I = \int_s J_z\,dy'$
M	$-J$	ρ	J_z

On the contrary, with the TEM duality, the sources remain the same retaining a microstrip configuration, but the substrate (ε_0, $\mu_0\mu_r$) is mapped to the already analyzed geometry with ($\varepsilon_0\varepsilon_r$, μ_0).

The TEM duality principle presented in Table 5.1 is built on Pucel and Masse [34, 35], but also on the original work used by them, which is that of Kaneki [38]. It is important to observe that TEM duality starts from the correspondence of $1/\varepsilon \to \mu$ and the charge density to the axial current density as $\rho \to J_z$. As pointed out by Kaneki [38], this duality yields the correspondence of the microstrip-line capacitance (C) in the dielectric substrate case to the reciprocal of the line inductance ($1/L$) in the magnetic substate case ($C \to 1/L$). In turn, the correspondence of

$\varepsilon_{reff} \to 1/\mu_{reff}$ stems from the definition of these effective quantities based on transmission line theory in conjuction with the waveguiding approach. This yields two alternative definitions for the phase velocity, as:

$$V_p = \frac{\omega}{\beta} = \frac{\omega}{\omega\sqrt{LC}} = \frac{1}{\sqrt{LC}}. \tag{5.53}$$

For a microstrip with air substrate ($\varepsilon_r = 1, \mu_r = 1$) assumed as a reference, let its inductance L_0 and capacitance C_0 and the phase constant $\beta = k_0$, then:

$$V_{p0} = \frac{\omega}{k_0} = C_{light} = \frac{1}{\sqrt{L_0 C_0}}. \tag{5.54}$$

In the dielectric substrate case ($\varepsilon_r, \mu_r = 1$), denote these quantities with a d-subscript (L_d, C_d) and the phase constant $\beta = k_0\sqrt{\varepsilon_{reff}}$, then:

$$V_{pd} = \frac{C_{light}}{\sqrt{\varepsilon_{reff}}} = \frac{1}{\sqrt{L_d C_d}}. \tag{5.55}$$

However, for this non-magnetic substrate, the inductance remains the same as in the air case ($L_d = L_0$):

$$\frac{V_{pd}}{V_{p0}} = \frac{1}{\sqrt{\varepsilon_{reff}}} = \frac{\sqrt{C_0}}{\sqrt{C_d}} \longleftrightarrow \varepsilon_{reff} = \frac{C_d}{C_0}. \tag{5.56}$$

Similarly, for the magnetic substrate ($\varepsilon_r = 1, \mu_r$), we let them as (L_m, C_m) and $\beta = k_0\sqrt{\mu_{reff}}$. So:

$$V_{pm} = \frac{C_{light}}{\sqrt{\mu_{reff}}} = \frac{1}{\sqrt{L_m C_m}}. \tag{5.57}$$

As this magnetic media has $\varepsilon_r = 1$, then the capacitance remains the same as for the air case ($C_m = C_0$), thus:

$$\frac{V_{pm}}{V_{p0}} = \frac{1}{\sqrt{\mu_{reff}}} = \frac{\sqrt{L_0}}{\sqrt{L_m}} \longleftrightarrow \mu_{reff} = \frac{L_m}{L_0}. \tag{5.58}$$

Hence, the relative effective permeability μ_{reff} can be estimated form the available effective dielectric constant $\varepsilon_{reff,d}$ as:

$$\mu_{reff}(w/h, \mu_r) = \frac{1}{\varepsilon_{reff,d}(w/h, \varepsilon_r \leftarrow 1/\mu_r)}. \tag{5.59}$$

For convenience, let us recall the approximate expressions of the effective dielectric constant ($\varepsilon_{reff,d}$) and the characteristic impedance (Z_{0d}) for the dielectric substrate, e.g., Pozar [1]:

$$\varepsilon_{reff,d} = \frac{\varepsilon_r + 1}{2} + \frac{\varepsilon_r - 1}{2} \frac{1}{\sqrt{1 + 12h/w}} \tag{5.60a}$$

$$Z_{0d} = \begin{cases} \frac{60}{\sqrt{\varepsilon_{reff,d}}} \ln\left(\frac{8h}{w} + \frac{w}{4h}\right) & \text{for } w/h \leq 1 \\ \frac{120}{\sqrt{\varepsilon_{reff,d}}\left[\frac{w}{h} + 1.393 + 0.667\ln\left(\frac{w}{h} + 1.444\right)\right]} & \text{for } w/h \geq 1. \end{cases} \tag{5.60b}$$

Alternatively, the effective medium can be characterized through the "filling factors" already used in the previous sections. For the single dielectric substrate, this can be written according to Wheeler [9] as:

$$q_d = \frac{\varepsilon_{reff} - 1}{\varepsilon_r - 1}. \tag{5.61}$$

In the presence of dielectric losses Pucel, et al. [39] improved the corresponding formula to account for loss tangent in the microstrip (only a dielectric slab rather than a dielectric filling of the entire space for $\tan \delta_d$), which is estimated as:

$$\tan \delta_{deff} = q_{d,l} \tan \delta_d, \tag{5.62}$$

where

$$q_{d,l} = q_d \frac{\varepsilon_r}{\varepsilon_{reff}} = \frac{1 - \varepsilon_{reff}^{-1}}{1 - \varepsilon_r^{-1}}. \tag{5.63}$$

Substituting for $\varepsilon_r \rightarrow 1/\mu_r$ and $\varepsilon_{reff,d} \rightarrow 1/\mu_{reff}$ in the above expressions (5.58) through (5.63) yields the desired quasi-TEM quantities for the $(\varepsilon_0, \mu_0\mu_r)$ substrate, as:

$$q_m\left(w/h, \mu_r\right) = q_d\left(w/h, \mu_r^{-1}\right) = \frac{\mu_{reff}^{-1} - 1}{\mu_r^{-1} - 1}, \tag{5.64}$$

where μ_{reff}^{-1} is given by Eqs.(5.59) and (5.60) or other more accurate expression for $\varepsilon_{reff,d}$ given for example in Hoffman's [3, 34]. Moreover, to account for magnetic losses ($\tan \delta_m$), the expression corresponding to Eqs. (5.62) and (5.63) read:

$$\tan \delta_{m,eff} = q_{m,l} \tan \delta_m \tag{5.65}$$

$$q_{m,l} = q_m \frac{1 - \mu_{reff}}{1 - \mu_r}. \tag{5.66}$$

Besides the above approach, our interest is for a ferrite substrate (even isotropic) which presents both dielectric and magnetic properties as $(\varepsilon_0\varepsilon_r, \mu_0\mu_r)$. According to [34, 35], its behavior can be considered as a superposition of the separate dielectric and magnetic substrates.

Thus, the effective index of refraction reads:

$$n_{eff} = \sqrt{\varepsilon_{reff,d} \cdot \mu_{reff}} \tag{5.67}$$

and the guided wavelength as:

$$\lambda_g = \frac{\lambda_0}{n_{eff}} = \frac{\lambda_0}{\sqrt{\varepsilon_{reff,d} \cdot \mu_{reff}}}. \tag{5.68}$$

Observing Eqs. (5.60), one may find out that the characteristic impedance of the dielectric substrate (Z_{0d}) is equal to that of air substrate $Z_0' = Z_{0d}(\varepsilon_r = 1, \mu_r = 1)$ divided by the effective dielectric constant [39], as:

$$Z_{0d} = \frac{Z_0'}{\sqrt{\varepsilon_{reff,d}}}. \tag{5.69}$$

Likewise, for the purely magnetic substrate ($\varepsilon_0, \mu_0\mu_r$), keeping in mind that $\varepsilon_{reff,d} \to 1/\mu_{reff}$, it is:

$$Z_{0m} = Z_0' \cdot \sqrt{\mu_{reff}}. \tag{5.70}$$

Hence, their superposition for the remnant magnetic ferrite substrate ($\varepsilon_0\varepsilon_r, \mu_0\mu_r$) yields:

$$Z_0 = Z_0' \cdot \sqrt{\mu_{reff}/\varepsilon_{reff,d}}. \tag{5.71}$$

Pucel and Masse [34, 35] give the attenuation factor (α_{dd}) due to the substrate media for the dielectric substrate (or Eq. (5.61)) as:

$$\alpha_{dd} \text{ (dB/cm)} = \frac{27.3}{\lambda_{gd}} \tan \delta_{deff}, \tag{5.72}$$

where $\lambda_{gd} = \lambda_0/\sqrt{\varepsilon_{reff,d}}$. Likewise, for the purely magnetic substrate using (5.65), it is:

$$\alpha_{dm} \text{ (dB/cm)} = \frac{27.3}{\lambda_{gm}} \tan \delta_{m,eff}, \tag{5.73}$$

where $\lambda_{gm} = \lambda_0/\sqrt{\mu_{reff}}$. In the combined case of remnant ferrite substrate, both media losses occur simultaneously, thus their superposition yields:

$$\alpha_d \text{ (dB/cm)} = \frac{27.3}{\lambda_g} \cdot \left[\tan \delta_{d,eff} + \tan \delta_{m,eff}\right], \tag{5.74}$$

where λ_g is defined in Eq. (5.68).

For an accurate expression of the ohmic conductor losses, the current density flowing up to the skin depth should be integrated according to the well-known formula derived from the Leontovich impedance boundary condition. In this case, there is a singularity at the strip edges where the current density parallel to the metallic edges or the field components normal to the edges tend to infinity at a rate $1/\sqrt{\rho}$, where ρ is the distance from the edge. However, for a simpler engineering formula, the current density could be approximated as uniform equal to the ratio of the total current (I) to the strip width w, (I/w) to yield an attenuation due to conductor losses [39] as:

$$\alpha_c \ (\text{dB/cm}) = 8.686 R_s/(Z_0 \cdot w), \tag{5.75}$$

where the conductor surface resistance reads:

$$R_s = \sqrt{\frac{\omega \mu_1}{2\sigma}} = \sqrt{\frac{\pi f \cdot \mu_1}{\sigma}}, \tag{5.76}$$

where μ_1 and σ are the strip conductor permittivity and conductivity. Usually, the metallic conductors are non-magnetic $\mu_1 = \mu_0 = 4\pi \cdot 10^{-7}$ H/m.

Returning to the ferrite substrate, assuming a DC magnetization along the propagation direction $\hat{z}$ as in Figure 5.8, the permeability tensor can be written as:

$$[\mu] = \mu_0 \begin{bmatrix} \mu_r & -jk_r & 0 \\ jk_r & \mu_r & 0 \\ 0 & 0 & \mu_{rz} \end{bmatrix}. \tag{5.77}$$

When the ferrite is magnetized to saturation, it is $\mu_{rz} \approx 1$, while this is different for partial magnetization. Equation (5.77) can be simplified for the two special cases to be considered herein.

(a) **Demagnetized substrate**

According to [35] and the references therein, especially Green et al. [40], in the demagnetized state, the off diagonal permeability tensor elements vanish, ($k_r = 0$), while the diagonal elements are equal to $\mu_r = \mu_{rdem}$. Since, the mode is quasi-TEM, the magnetic field in the propagation direction is negligible, ($Hz \approx 0$). Hence, the demagnetized ferrite substrate can be considered as the isotropic case ($\varepsilon_0 \varepsilon_r$, $\mu_0 \mu_{rdem}$) analyzed above, where [41]:

$$\mu_{rdem} \approx \frac{1}{3} \left\{ 1 + 2\sqrt{1 - \left(\frac{\omega_m}{\omega}\right)^2} \right\}, \tag{5.78}$$

where $\omega_m = \gamma \mu_0 M_s$ or equal to $\gamma \cdot (4\pi M_s)$ in CGS units.

(b) **Latched substrate at remnant magnetization**

Adapting the analysis of [35] as well as the original work of Green et al. [41], the phenomenological expression for μ_r, k_r, μ_{rz} of permeability tensor elements for a partially magnetized ferrite at $M < M_s$ reads:

$$\mu_r = \mu_{rdem} + (1 - \mu_{rdem}) \cdot \left(\frac{4\pi M}{4\pi M_s} \right)^{3/2} \tag{5.79}$$

$$k_r = \frac{\gamma \cdot (4\pi M)}{\omega} \tag{5.80}$$

$$\mu_{rz} = \mu_{rdem} \cdot [1 - (4\pi M/4\pi M_s)]^{5/2} . \tag{5.81}$$

When the ferrite substrate is latched to remanence ($M = M_r$), the magnetization becomes $4\pi M = 4\pi M_r$.

However, the substrate permeability remains anisotropic (tensor), while the above analysis encounters only isotropic magnetic substrates. Sandy and Sage [42], based on a finite difference approach, proposed *an equivalent isotropic permeability* μ_{rmag} by curve fitting the numerical data obtained from the actual gyrotropic media. Masse and Pucel experimentally validated their approximation in [35], and this reads:

$$\mu_{rmag} = \frac{\mu_r^2 - k_r^2}{\mu_r} \frac{1}{1 - \frac{1}{7}\sqrt{\frac{h}{w}} \left(\frac{k_r}{\mu_r} \right)^2 \ln \left(1 + \frac{\mu_r}{\mu_r^2 - k_r^2} \right)} , \tag{5.82}$$

where μ_r and k_r are given in Eqs. (5.79) and (5.80). Even though Sandy and Sage [42], did not claimed physical explanation for (5.82) it is very interesting to observe that their expression involves the effective permeability of "*transversely magnetized ferrite specimens*," namely, $\mu_{reff} = (\mu_r^2 - k_r^2)/\mu_r$.

Summarizing, μ_{rmag} along with the ferrite dielectric constant ε_r can be used in the previous analysis to yield the characteristics of a microstrip line printed on a partially (weakly) magnetized ferrite.

5.15   MICROSTRIP ON A SINGLE ANISOTROPIC DIELECTRIC SUBSTRATE

In many practical cases tunable materials like ferrites or solid-state plasmas present gyrotropic tensor constitutive parameters for certain DC magnetic bias. These present anisotropic permeability ($\varepsilon, [\mu]$) and permittivity ($[\varepsilon], \mu$), respectively. The gyrotropic tensor parameters can be written in general form as:

$$[\varepsilon] = \begin{bmatrix} \varepsilon_{xx} & \varepsilon_{xy} & 0 \\ \varepsilon_{yx} & \varepsilon_{yy} & 0 \\ 0 & 0 & \varepsilon_{zz} \end{bmatrix} , \qquad [\mu] = \begin{bmatrix} \mu_{xx} & \mu_{xy} & 0 \\ \mu_{yx} & \mu_{yy} & 0 \\ 0 & 0 & \mu_{zz} \end{bmatrix} . \tag{5.83}$$

The same permittivity tensor describes many crystalline dielectric substrates like sapphire, and their analysis attracted significant research interest. Based on these similarities and the *TEM mode duality principle* described in the previous section, it seems a good idea to review the techniques already established for anisotropic crystalline substrates, which will then be exploited for the analysis of tunable substrates. An extensive review of the literature until 1985 is given by Alexopoulos [43], covering both approximate-empirical methods as well as full-wave numerical techniques. Herein, only the approximate-empirical methods are recalled, since they may serve the engineering design approach.

(a) **Equivalent isotropic media for sapphire substrate**

The crystalline sapphire substrate presents a diagonalized permittivity tensor as $\varepsilon_{rxx} = \varepsilon_{rzz} = \varepsilon_{r\perp} = 9.34$ and $\varepsilon_{ryy} = \varepsilon_{r//} = 11.49$, where its optical axis is assumed aligned along the $\hat{y}$-axis (by appropriate crystal cut), which is perpendicular to the microstrip substrate (Figure 5.10). Owens et al. [44], proposed an equivalent isotropic substrate after an extensive analysis of the dominant quasi-TEM mode of microstrip lines printed on sapphire substrate using the finite difference method as:

$$\varepsilon_{req} = 12.0 - 1.21 / \left\{ 1 + 0.39 \left[\log \left(10w/h \right) \right]^2 \right\}. \tag{5.84}$$

The accuracy of Eq. (5.84) is estimated as being $\pm 0.5\%$ in the range $0.1 \leq w/h \leq 10$. *The effective dielectric constant and the characteristic impedance (Z_0) can be estimated using the formulas of the ordinary microstrip lines printed on isotropic substrate.*

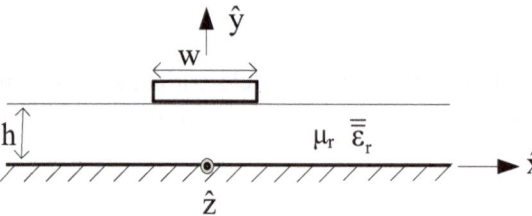

Figure 5.10: Microstrip line printed on a uniaxial anisotropic substrate with the optical axis transversely aligned along $\hat{y}$.

(b) **Microstrip on anisotropic substrate with optical axis misalignment**

Dielectric crystalline substrates (even like sapphire) may present non-zero off diagonal elements in their permittivity tensor depending on how the crystal is cut to form the substrate. In practice, this process usually yields an optical axis misalignment which forms an angle θ with respect to the y-axis (in the transverse xy plane as in Figure 5.11) instead of being perfectly aligned along the $\hat{y}$-axis. In this case, the dielectric constant tensor takes a form "similar" to a gyroelectric media as given in Eq. (5.83). Considering [43] or [45], with the notation of Figure 5.11, we have:

$$\varepsilon_{xx} = \varepsilon_{\xi\xi} \cos^2 \theta + \varepsilon_{\eta\eta} \sin^2 \theta \tag{5.85a}$$

$$\varepsilon_{yy} = \varepsilon_{\xi\xi} \sin^2 \theta + \varepsilon_{\eta\eta} \cos^2 \theta \tag{5.85b}$$

$$\varepsilon_{xy} = \varepsilon_{yx} = \left(\varepsilon_{\xi\xi} - \varepsilon_{\eta\eta} \right) \sin \theta \cos \theta \tag{5.85c}$$

$$\varepsilon_{zz} = \varepsilon_{\zeta\zeta}, \tag{5.85d}$$

where $\varepsilon_{\xi\xi}$ and $\varepsilon_{\eta\eta}$ are the dielectric constants along the principal axes (ξ, η, ζ) of the crystal. The permittivity tensor becomes diagonal in that coordinate system. Note that ε_{yy} is corrected according to Yu and Rawat [46].

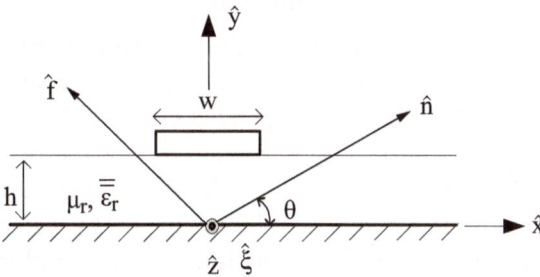

Figure 5.11: Microstrip on a uniaxial anisotropic substrate with possible optical axis misalignment by an angle θ [43].

(c) Coordinate transformation method

The above anisotropic substrate can be transformed to an equivalent isotropic one with equivalent dielectric constant ε_{req} and equivalent thickness h_{eq}, according to Alexopoulos [43], as:

$$\varepsilon_{req} = \left[\varepsilon_{rxx} \cdot \varepsilon_{ryy} - \varepsilon_{rxy}^2 \right]^{1/2} \tag{5.86a}$$

$$h_{eq} = h \cdot \left[\frac{\varepsilon_{rxx}}{\varepsilon_{ryy}} - \left(\frac{\varepsilon_{rxy}}{\varepsilon_{ryy}} \right)^2 \right]^{1/2}. \tag{5.86b}$$

The validity of such a coordinate transformation has been rigorously proved for the diagonal case ($\varepsilon_{xy} = \varepsilon_{yx} = 0$) in [47, 48] by letting $(x, y) \to (\tau, \upsilon)$ as:

$$\tau = x \quad \text{and} \quad \upsilon = y \cdot \sqrt{\varepsilon_{rxx}/\varepsilon_{ryy}} \tag{5.87}$$

resulting to an equivalent isotropic substrate:

$$\varepsilon_{req} = \sqrt{\varepsilon_{rxx} \cdot \varepsilon_{ryy}} \tag{5.88a}$$

$$h_{eq} = h \cdot \sqrt{\varepsilon_{rxx}/\varepsilon_{ryy}}. \tag{5.88b}$$

As noted by Alexopoulos [43], the above equivalent isotropic substrate approach should be used with caution for microstrips with cover layer (shielded), since there is an indication that this concept may not be valid.

(d) **Frequency dispersion for anisotropic substrates**

The equivalent isotropic substrate formulas given in Eqs. (5.86) and (5.87) are obtained through quasi-static methods and are exclusively applicable to sapphire substrates. Hence, they do not account for the frequency dispersion of the effective dielectric constant and need to be extended to an arbitrary substrate anisotropy, but of the form of Eq. (5.83). Obviously, one could adopt the classical formulas accounting for frequency dispersion of isotropic substrates as given for example in Hoffman's [3] by just letting $\varepsilon_{req} \to \varepsilon_r$ and $h_{eq} \to h$. An empirical formula established specifically for the sapphire substrate by Eduards and Owens in [49] reads:

$$\text{Sapphire substrate: } \varepsilon_{reff} = \varepsilon_{req} - \frac{\varepsilon_{req} - \varepsilon_{reff0}}{1 + (h_m/Z_0)^{1.33} \left(0.43 f_g^2 - 0.009 f_g^3\right)}, \tag{5.89}$$

where ε_{req} is given in (5.84). ε_{reff0} is the static effective dielectric constant and $h_m = h$ (mm), $f_g = f$ (GHz), also according to [3]. Note that ε_{reff0} is also given in the previous section of this chapter as $\varepsilon_{reff}(f = 0)$.

(e) **Arbitrary anisotropic substrate frequency dispersion**

For an arbitrary anisotropic dielectric substrate, the equivalent isotropic characteristics (ε_{req} and h_{eq}) are given in Eqs. (5.86). Once again, the frequency dispersion regarding ε_{reff} and Z_0 can be obtained from the well-established isotropic substrate formulas, e.g. [3]. For convenience, the Kobayashi formula [50], appearing also in [43], is repeated herein for completeness and convenience as:

$$\varepsilon_{reff} = \begin{cases} \frac{4}{(I_1+I_2)^2} & \text{if } I_1 > I_2 \\ \frac{1}{I_2^2} & \text{if } I_1 < I_2. \end{cases} \tag{5.90}$$

The quantities I_1, I_2 are the isotropic substrate dispersive models established by Schneider [51] and Yamashita [52] as:

$$I_1 = \left\{ \frac{1}{\sqrt{\varepsilon_{yy}}} \left(\frac{f}{f_k} \right)^2 + \frac{1}{\sqrt{\varepsilon_{eff0}}} \right\} \bigg/ \left\{ \left(\frac{f}{f_k} \right)^2 + 1 \right\} \qquad (5.91)$$

with

$$f_k = \frac{v_0 \tan^{-1} \left[\varepsilon_{req} \left(\frac{\varepsilon_{eff0}-1}{\varepsilon_{req}-\varepsilon_{eff0}} \right)^{1/2} \right]}{2\pi H \left(1 + w/H \right) \left(\varepsilon_{req} - \varepsilon_{eff0} \right)^{1/2}} \qquad (5.92)$$

and

$$I_2 = \left\{ \left(\frac{f}{f_y} \right)^{3/2} + 4 \right\} \bigg/ \left\{ \left(\frac{f}{f_y} \right)^{3/2} \sqrt{\varepsilon_{yy}} + 4\sqrt{\varepsilon_{eff0}} \right\} \qquad (5.93)$$

with

$$f_y = \frac{v_0}{4H \sqrt{\varepsilon_{req} - 1} \left\{ \frac{1}{2} + \left[1 + 2 \log \left(1 + \frac{w}{H} \right) \right]^2 \right\}}. \qquad (5.94)$$

In the above expressions, ε_{req} is defined by Eq. (5.86a). $H = h$ (mm) and the quantities (w, h) are those of the original line, not the equivalent ones. Also, $v_0 = C$ is the speed of light in vacuum. Details regarding the accuracy of (5.90) are given in [43] and the references therein.

5.16 MICROSTRIP PRINTED ON A WEAKLY MAGNETIZED FERRITE–DIELECTRIC SUBSTRATE

The fabrication of microstrip structures directly on a ferrite substrate (e.g., phase shifters) involves numerous difficulties regarding the metallization of a ferrite surface. Printing the microstrip structure on an ordinary dielectric substrate (appropriate for microwave frequencies) and overlaying this on top of a grounded ferrite slab as shown in Figure 5.12 constitutes a convenient practice [53]. Additionally, the dielectric substrates reduce the losses (decreasing the attenuation constant) and improve the power-handling capability. However, it entails a reduction between the current wave flowing on the strip conductor and the electromagnetic fields inside the ferrite, resulting in an undesired reduction in phase shift efficiency.

Even though the actual interest in this structure concerns the case when the ferrite is magnetized to its saturation ($M = M_S$), herein the simpler case of its partial magnetization will be considered in order for this structure to be handled analytically. Thus, the ferrite is considered *partially magnetized* by a DC magnetic field aligned along the propagation axis-$\hat{z}$. Hence, its tensor permeability is given by Eqs. (5.77) and (5.79)–(5.81). From a first point of view, an equivalent isotropic substrate approach can be employed to yield an approximate scalar relative

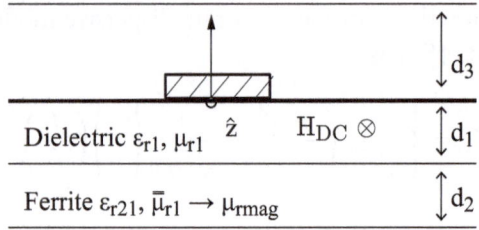

Figure 5.12: Microstrip line over a composite dielectric-weakly magnetized ferrite.

permeability μ_{rmag} given by Eq. (5.82). However, Bolioli et al. [53] retained only the first term of (5.82) as the equivalent magnetic permeability:

$$\mu_{req} = \mu_{rT} = \frac{\mu_r^2 - k_r^2}{\mu_r}. \tag{5.95}$$

In turn, the same authors considered the Pucel and Masse [34] duality principle $\varepsilon_r \leftarrow 1/\mu_{rT}$ to establish an effective permeability. Their procedure for the composite ferrite-dielectric substrate is illustrated in Figure 5.13, where the quasi-static effective permeability of a dual dielectric layer (Figure 5.13a) is obtained from the related capacitances as usual:

$$\varepsilon_{reff} = \frac{C}{C_0} = \frac{\text{strip capacitance of dual dielectric substrate } (d_1, d_2)}{\text{strip capacitance of an air substrate } (d_1 + d_2)}. \tag{5.96a}$$

By adopting the TEM mode duality principle $\varepsilon_r \leftarrow 1/\mu_r$ as in Figure 5.13b, the $1/\mu_{reff}$ for a substrate with thickness $d_1 + d_2$ as in Figure 5.13c is:

$$1/\mu_{reff} = C_1/C_0 \leftrightarrow \mu_{reff} = C_0/C_1. \tag{5.96b}$$

Following the Pucel and Masse [34] approach given in the previous section (see Eq. (5.48)) the propagation constant and the characteristic impedance are defined as illustrated in Figure 5.14 and given as:

$$Z_0 = \frac{\sqrt{\varepsilon_{reff} \cdot \mu_{reff}}}{C_{light} \cdot C} = \frac{1}{v_p \cdot C} \tag{5.97}$$

$$\beta = \frac{2\pi f}{C_{light}} \cdot \sqrt{\varepsilon_{reff} \cdot \mu_{reff}} = k_0 \cdot \sqrt{\varepsilon_{reff} \cdot \mu_{reff}} = \frac{2\pi}{\lambda_g}, \tag{5.98}$$

where C_{light} is the speed of light in free space and C the total strip capacitance for the composite substrate.

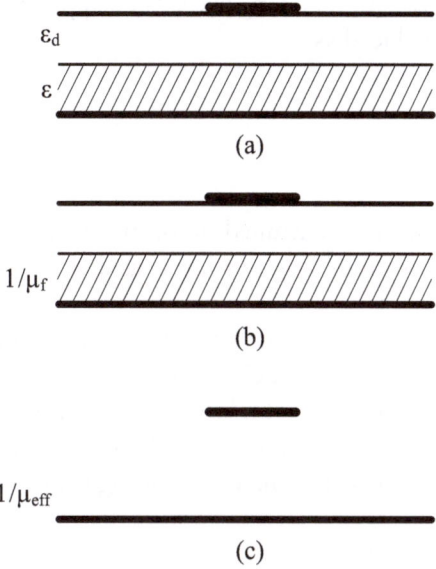

ε_d

ε

(a)

$1/\mu_f$

(b)

$1/\mu_{eff}$

(c)

Figure 5.13: Definition of the effective permeability of a ferrite-dielectric composite, as in [53].

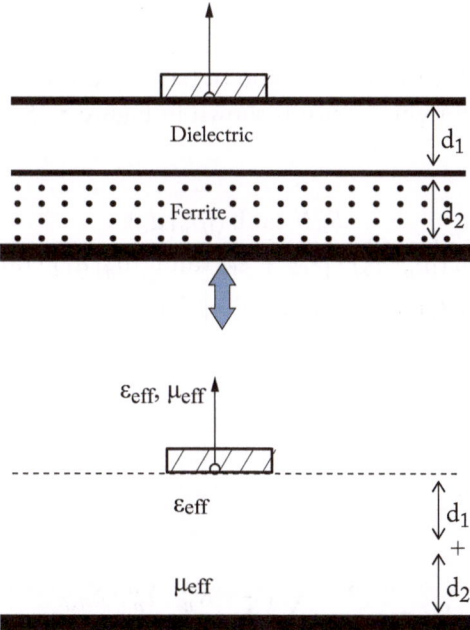

Dielectric

d_1

Ferrite

d_2

$\varepsilon_{eff}, \mu_{eff}$

ε_{eff}

d_1

$+$

μ_{eff}

d_2

Figure 5.14: Definition of effective electric permittivity and effective magnetic permeability for a ferrite-dielectric composite [53].

Besides the above expression, following the simple approximation given in [3], the characteristic impedance can be deduced as:

$$Z_0(f \to 0) = \frac{Z_{0,air}}{\sqrt{\varepsilon_{reff}\mu_{reff}}}. \tag{5.99}$$

The effective permittivity is determined according to (5.96) by ignoring the magnetic properties of the ferrite, e.g., assuming the two layers as dielectrics with $(\varepsilon_0\varepsilon_{rd}, \mu_0)$ and $(\varepsilon_{rF}\varepsilon_0, \mu_0)$.

Observing Eqs. (5.97)–(5.99), the characteristic of the microstrip line can be determined through the evaluation of capacitances C_0, C, and C_1. For this purpose, Bolioli et al. [53] employed a variational and a least-squares boundary residual (LSBR) method in the spectral domain for an electrostatic TEM approximation to estimate ε_{reff} and a magnetostatic TEM approximation to evaluate μ_{reff} in closed form. The resulting expressions are as follows [53]:

Electrostatic (for ε_{reff}):

$$C = \frac{Q}{V} \quad \text{and} \quad V = \frac{2}{b\sqrt{\delta}} \cdot \sum_n \sum_p \frac{K_n^e \cdot Y_n^e \cdot S_{n,p}}{n \cdot \beta'}, \tag{5.100}$$

where b is the substrate width and δ is the ratio of (strip width w)/(number of basis functions) representing the unknown current density as shown in Figure 5.15.

$$K_n^e = \frac{1}{\varepsilon_0} \frac{S_n N \tanh\left(|\beta_n'|\, d_3\right)}{N + \varepsilon_{r1} \tanh\left(|\beta_n'|\, d_3\right) \cdot \left[\varepsilon_{r2} + \varepsilon_{r1} \tanh\left(|\beta_n'|\, d_1\right) \cdot \tanh\left(|\beta_n'|\, d_2\right)\right]} \tag{5.101a}$$

$$N = \varepsilon_{r2} \cdot \tanh\left(|\beta'n|\, d_1\right) + \varepsilon_{r1} \cdot \tanh\left(|\beta'n|\, d_2\right) \tag{5.101b}$$

$$S_n = |n|/n \quad \text{and} \quad \beta' = 2\pi/b. \tag{5.101c}$$

Magnetostatic (for $1/\mu_{reff}$):

$$C_1 = \frac{Q}{V_m} \quad \text{and} \quad V_m = \frac{2}{b\sqrt{\delta}} \cdot \sum_n \sum_p \frac{K_n^h \cdot Y_n^h \cdot S_{n,p}}{n \cdot \beta}, \tag{5.102}$$

$$K_n^h = \frac{S_n \cdot N \cdot \tanh\left(|\beta'n|\, d_3\right)}{\tanh\left(|\beta'n|\, d_1\right) + \tanh\left(|\beta'n|\, d_3\right)} \cdot \frac{N_1 + N_2 \cdot \tanh\left(|\beta'n|\, d_2\right)}{D_1 + D_2 \cdot \tanh\left(|\beta'n|\, d_3\right)}, \tag{5.103}$$

where N, S_n, β' are the same as in (5.101), while additionally:

$$D_1 = \mu_0 \mu_r \left|\beta'n\right|^2 \tag{5.104a}$$

$$N_1 = D_1 \cdot \tanh\left(\left|\beta'n\right| d_1\right) \tag{5.104b}$$

$$D_2 = \frac{\mu_0 \left(\mu_r^2 - k_r^2\right)\left(\beta'_n\right)^2}{\tanh\left[\left|\beta'_n\right| \cdot (d_1 + d_3)\right]} - \mu_0 k_r \beta'_n \left|\beta'_n\right| \tag{5.104c}$$

$$N_2 = \mu_0 \left(\mu_r^2 - k_r^2\right)\left(\beta'_n\right)^2 - \mu_0 k_r \beta'_n \left|\beta'_n\right| \tanh\left(\left|\beta'_n\right| d_1\right). \tag{5.104d}$$

The summation over-n concerns the discretized charge density over the strip as shown in Figure 5.15.

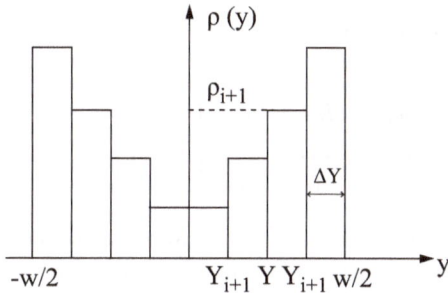

Figure 5.15: Discretization of the charge density over the strip conductor [53].

Bolioli et al. [53] exploited the above theory for the design and implementation of phase shifters.

(a) **Application to phase shifters**

The phase shifters implemented in [53] utilized an Alumina dielectric layer (Al_2O_3) with $\varepsilon_r = 9.6$, thickness $d_1 = 0.254$ mm and two alternative Thomson/CSF ferrites of types (A230, thickness $d_2 = 1.15$ mm) and (A28, thickness $d_2 = 0.8$ mm). Both of them have the same dielectric constant $\varepsilon_{r2} = 16.6$, while their saturation magnetization is $4\pi M_S = 2300$ gauss for the A230 and $4\pi M_S = 2800$ gauss for A28. In the A230 case, the strip width was calculated as $w = 0.9$ mm to give a characteristic impedance $Z_0 = 50\,\Omega$ when operated at 10 GHz. Respectively, a width of $w = 0.67$ mm gave $Z_0 = 50\,\Omega$ at 9 GHz for A28 case. The DC magnetic bias in both cases is created using a 2,000-turn solenoid wrapped around the phase shifter (microstrip line) which was supplied by a varying DC-current up to 100 mA for A230 and up to 500 mA for A28, resulting in an almost linear variation in the phase shift. Beyond this range, a nonlinear response appears and the corresponding maximum magnetizations were $4\pi M \rightarrow 4\pi M_{\text{max}}$ are 1,740 gauss and 2,240 gauss, respectively.

5.17   MICROSTRIP LINES ON GYROTROPIC SUBSTRATE

Magnetized ferrite or semiconductor substrates present the so-called *gyrotropic* permeability ($[\mu]$) and permittivity ($[\varepsilon]$) tensors, when their DC-biasing magnetic field is aligned along the propagation direction axis:

Gyromagnetic, Polder [54]:

$$\varepsilon = \varepsilon_0 \cdot \varepsilon_r, \quad [\mu] = \mu_0 \begin{bmatrix} \mu_{11} & j\mu_{12} & 0 \\ -j\mu_{12} & \mu_{11} & 0 \\ 0 & 0 & \mu_{33} \end{bmatrix} = \mu_0 \begin{bmatrix} \mu_r & jk_r & 0 \\ -jk_r & \mu_r & 0 \\ 0 & 0 & \mu_{33} \end{bmatrix}. \quad (5.105)$$

Gyroelectric, Palik and Furdyna [55]:

$$\mu = \mu_0, \quad [\varepsilon] = \varepsilon_0 \begin{bmatrix} \varepsilon_{11} & j\varepsilon_{12} & 0 \\ -j\varepsilon_{12} & \varepsilon_{11} & 0 \\ 0 & 0 & \varepsilon_{33} \end{bmatrix} = \varepsilon_0 \begin{bmatrix} \varepsilon_r & jg_r & 0 \\ -jg_r & \varepsilon_r & 0 \\ 0 & 0 & \varepsilon_{33} \end{bmatrix}. \quad (5.106)$$

When magnetized to saturation ($4\pi M \rightarrow 4\pi M_S$), the dynamic field components and the corresponding effects of the constitutive relations μ_{33} and ε_{33} become negligible along the DC biasing direction, hence $\mu_{33} \rightarrow 1$ and $\varepsilon_{33} \rightarrow 1$.

It is by now well established that magnetized semiconductors behave similarly to cold magnetoplasma, e.g., [55], at cryogenic temperatures and is widely known as "solid state plasma." With the recent introduction of superconductors like YBCO that operate at similar cryogenic conditions, solid-state plasma is becoming very important since it allows the implementation of "non-reciprocal" microwave devices (isolators, circulators) on the semiconductor substrate. Burneika et al. extensively examine microstrip lines printed on a magnetized semiconductor as well as on magnetized ferrite-semiconductor layers in [56] and the original references cited therein. In [57], a longitudinally magnetized (along the propagation $\hat{z}$-direction) n-type InSb operating at a cryogenic temperature of 77 K established by liquid nitrogen has a gyroelectric behavior as Eq. (5.106) with $\varepsilon_{11} = 19.6 + j1.3$ and $\varepsilon_{12} = -20.7 - j0.2$. The DC biasing magnetic field is $B_{DC} = 0.4$ tesla.

A ferrite layer under the same DC bias and the same operating conditions presents gyromagnetic behavior as in (5.105) with $4\pi M_S = \gamma M_S = 369.8$ KA/m and $\mu_{11} = 0.97 + j0.0027$, $\mu_{12} = -0.21 - j0.0084$, and $\varepsilon_r = \varepsilon_{r,fer} = 10$. A first significant observation is that the imaginary parts of ε_{11} and ε_{12} are quite larger as compared to the imaginary parts of μ_{11} and μ_{12} (even taken as $\tan \delta_\varepsilon = \text{Im}(\varepsilon)/\text{Re}(\varepsilon)$ vs. $\tan \delta_\mu = \text{Im}(\mu)/\text{Re}(\mu)$). This means that the solid-state plasma losses are at least 5–10 times larger than the ferrite losses.

5.18   TEM DUALITY PRINCIPLE IN GYROTROPIC MEDIA

The authors in [56], elaborating on a TEM quasi-static analysis of gyrotropic media, were able to establish a very important TEM duality principle in gyroelectric and gyromagnetic substrates by exploiting a "singular integral equation" methodology for static problems. Explicitly they proved that the boundary conditions of a microstrip line printed on a gyroelectric substrate and operated at its dominant quasi-TEM mode (electrostatic potential) become identical to the boundary conditions of the magnetostatic analysis when the following substitution is adopted:

$$\varepsilon_{11} \to \mu_{11}^{-1} \quad \text{and} \quad \varepsilon_{12} \to -\mu_{12}^{-1}. \tag{5.107}$$

Under the same substitution, the strip electrostatic charge density (ρ) becomes identical to the surface current density (j_s) in the magnetostatic case (see Eqs. (5-13) and (5-14) in [56]). This is actually a rigorous proof of the *TEM duality principle between gyroelectric and gyromagnetic substrates*. Besides being a generalization of Pucel et al. [39], TEM duality principle $\varepsilon_r \to 1/\mu_r$ between the scalar (isotropic) dielectric substrates offers the required self-consistency.

Additionally, the analysis in [56] provides some useful formulas for the transverse wavenumbers for the general case of wave propagation in gyrotropic waveguides, which are valid for any mode type TE_Z, TM_Z or HE_Z, EH_Z as well as the herein desired quasi-TEM mode. These transverse wavenumbers are expressed in [56] as:

$$k_{\perp 1,2}^2 = \frac{1}{2} \left\{ \begin{array}{l} k^2 \left(\varepsilon_{r\perp} + \mu_{r\perp}\right) - h^2 \left(\frac{1}{\varepsilon_{11}} + \frac{1}{\mu_{11}}\right) \pm \left[k^2 \left(\varepsilon_{r\perp} - \mu_{r\perp}\right) + h^2 \left(\frac{1}{\varepsilon_{11}} - \frac{1}{\mu_{11}}\right)\right]^2 + \\ 4h^2 k^2 \left(\frac{\varepsilon_{12}}{\varepsilon_{11}} + \frac{\mu_{12}}{\mu_{11}}\right)^2 \end{array} \right\}, \tag{5.108}$$

where $k = k_0$ the free-space wavenumber, $h = \beta$ the axial (longitudinal) propagation constant and

$$\varepsilon_{r\perp} = \frac{\varepsilon_{11}^2 - \varepsilon_{12}^2}{\varepsilon_{11}}, \qquad \mu_{r\perp} = \frac{\mu_{11}^2 - \mu_{12}^2}{\mu_{11}}. \tag{5.109}$$

For the case of gyroelectric medium ($\varepsilon_0[\varepsilon_r]$, $\mu_0\mu_r$), we may set $\mu_{11} = \mu_r$ and $\mu_{12} = 0$. Likewise, for a gyromagnetic media ($\varepsilon_0\varepsilon_r$, $\mu_0[\mu_r]$), one may set $\varepsilon_{11} = \varepsilon_r$ and $\varepsilon_{12} = 0$. For a non-gyrotropic medium setting $\varepsilon_{12} = 0$, $\mu_{12} = 0$, Eq. (5.108) yields:

$$k_{\perp 1}^2 = k^2 \frac{\varepsilon_{11}\mu_{11} - h^2}{\mu_{11}} \tag{5.110a}$$

$$k_{\perp 2}^2 = k^2 \frac{\varepsilon_{11}\mu_{11} - h^2}{\varepsilon_{11}}. \tag{5.110b}$$

Equations (5.110) represent pure TE_Z and TM_Z wavenumbers.

5.19   REFERENCES

[1] D. M. Pozar, *Microwave Engineering*, 4th ed., Wiley 2012. DOI: 10.1063/1.2836410. 179, 192, 202, 208

[2] R. E. Collin, *Foundations for Microwave Engineering*, Int. ed., McGraw-Hill, 1992. DOI: 10.1146/annurev-matsci-070813-113315. 179

[3] R. K. Hoffman, *Handbook of Microwave Integrated Circuits*, Artech House, 1987. DOI: 10.1002/mmce.20961. 179, 180, 183, 187, 188, 190, 191, 192, 208, 214, 218

[4] K. C. Cupta, R. Garg, I. Bahl, and P. Bhartia, *Microstrip Lines and Slotlines*, 2nd ed., Artech House 1996. DOI: 10.1088/0953-8984/20/43/434220. 179, 193, 199, 202, 203

[5] E. Yamashita, Variational method for the analysis of microstrip-like transmission lines, *IEEE Transactions on Microwave Theory and Techniques*, vol. MTT-16, pp. 529–535, August 1968. DOI: 10.1016/0025-5408(78)90161-7. 179, 183, 191

[6] A. Farrar and A. T. Adams, Multilayer microstrip transmission lines, *IEEE Transactions on Microwave Theory and Techniques*, vol. MTT-22, pp. 889–891, October 1974. 179, 183

[7] J. Svacina, Analysis of multilayer microstrip transmission lines by conformal mapping method, *IEEE Transactions on Microwave Theory and Techniques*, vol. MTT-40, pp. 769–772, April 1992. xvi, 180, 182, 183, 189

[8] H. A. Wheeler, Transmission line properties of parallel wide strips by a conformal mapping approximation, *IEEE Transactions on Microwave Theory and Techniques*, vol. MTT-12, pp. 280–289, March 1964. 180

[9] H. A. Wheeler, Transmission line properties of parallel strips separated by a dielectric sheet, *IEEE Transactions on Microwave Theory and Technique*, vol. MTT-13, pp. 172–185, March 1965. 180, 188, 189, 190, 205, 208

[10] B. E. Spielman, Dissipation loss effects in isolated and coupled transmission lines, *IEEE Transactions on Microwave Theory and Techniques*, vol. MTT-25, p. 648, August 1977. 183

[11] P. Pramanick and P. Bhartia, Computer-aided design models for millimeter-wave finlines and suspended-substrate microstrip lines, *IEEE Transactions on Microwave Theory and Techniques*, vol. MTT-33, p.1429, December 1985. 183

[12] J. Svacina, A simple quasi-static determination of basic parameters of multilayer microstrip and coplanar waveguide, *IEEE Microwave Guided Wave Letters*, vol. 2, pp. 385–387, October 1992. xvi, 184, 187, 195, 199, 200

[13] Wan and A. Hoorfar, Improved design equations for multilayer microstrip lines, *IEEE Microwave and Guided Wave Letters*, vol.10, pp. 223–224, June 2000. 187

[14] E. Hammerstad and O. Jensen, Accurate models for microstrip computer-aided design, *IEEE MTT-S International Symposium Digest*, Washington DC, May 1980. 187

[15] K. Verma and G. Hassani Sadr, Unified dispersion model for multilayer microstrip line, *IEEE Transactions on Microwave Theory and Techniques*, vol. MTT-40, pp. 1587–1591, July 1992. 187, 188, 190, 191

[16] M. Kirschning and R. H. Jansen, Accurate model for effective dielectric constant of microstrip with validity up to millimetre-wave frequencies, *Electronics Letters*, vol. 18, no. 6, pp. 272–3, March 1982. 187, 190, 191

[17] A. K. Verma and A. Bhupal, Analysis of multilayer microstrip line slow-wave structure using single-layer reduction (SLR) formalism, *Microwave and Optical Technology Letters*, vol. 19, no. 5, pp. 371–375, December 1998. 188, 191

[18] A. K. Verma and A. Bhupal, Dielectric loss of multilayer microstrip line, *Microwave and Optical Technology Letters*, vol. 17, no. 6, pp. 368–370, April 1998. 188, 191

[19] R. Crampagne, M. Ahmadpanah, and J. L. Guiraud, A simple method for determining the green's function for a large class of MIC lines having multilayered dielectric structures, *IEEE Transactions on Microwave Theory and Techniques*, vol. MTT-26, pp. 82–87, February 1978. 188, 191

[20] M. Kobayashi, A dispersion formula satisfying recent requirements in microstrip CAD, *IEEE Transactions on Microwave Theory and Techniques*, vol. MTT-36, pp. 1246–1250, 1988. 190, 191

[21] M. L. Tounsi, M. C. E. Yagoub, and B. Haraoubia, New design formulas for microstrip transmission lines using high-dielectric substrate, *COMPEL: International Journal of Computation and Mathematics in Electrical and Electronic Engineering*, vol. 24, no. 1, pp. 15–34, 2005. 191

[22] F. C. De Ronde, A new class of microstrip directional couplers, *Digest of Technical Papers*, pp. 185–189, G-MTT Symposium, 1970. 193

[23] C. P. Wen, Coplanar waveguide: A surface strip transmission line suitable for non-reciprocal gyromagnetic device applications, *IEEE Transactions on Microwave Theory and Techniques*, vol. MTT-17, pp. 1087–1090, 1969. 193

[24] S. Gevorgian, H. Berg, H. Jacobsson, and T. Lewin, Basic parameters of coplanar-strip waveguides on multilayer dielectric/semiconductor substrates, part 1: High permittivity substrates, *IEEE Microwave Magazine*, pp. 60–70, June 2003. 193, 199

[25] M. Abramowitz and I. A. Stegun, *Handbook of Mathematical Functions*, 10th ed., Dover Publishers, NY, 1972. 195, 202

[26] I. S. Gradshteyn, I. M. Ryzhik, Yu V. Geronimus, M. Yu Tseytlin, and A. Jeffrey, *Table of Integrals, Series and Products*, Academic Press, NY, 1965. 198, 199

[27] G. Arfken, *Mathematical Methods for Physicists*, 2nd ed., Academic Press, NY, 1970.

[28] W. Hilberg, From approximations to exact relations for characteristic impedances, *IEEE Transactions on Microwave Theory and Techniques*, vol. MTT-17, pp. 259–265, 1969. 199

[29] G. Ghione and C. Naldi, Analytical formulas for coplanar lines in hybrid and monolithic MICs, *Electronics Letters*, vol. 20, no. 4, pp. 179–181, February 1984. 200

[30] E. Chen and S. Y. Chou, Characteristics of coplanar transmission lines on multilayer substrates: Modelling and experiments, *IEEE Transactions on Microwave Theory and Techniques*, vol. MTT-45, pp. 939–945, June 1997. 200, 202, 203

[31] C. Veyres and V. Fouad Hanna, Extension of the application of conformal mapping techniques to coplanar lines with finite dimensions, *International Journal of Electronics*, vol. 48, pp. 47–56, 1980. 200

[32] V. Fouad Hanna and D. Thebault, Theoretical and experimental investigation extension of asymetric coplanar waveguides, *IEEE Transactions on Microwave Theory and Techniques*, vol. MTT-32, pp. 1649–1651, December 1984. 203

[33] R. E. Collin, *Field Theory of Guided Waves*, 2nd ed., IEEE Press, NY, 1991. 204

[34] R. A. Pucel and D. J. Masse, Microstrip propagation on magnetic substrates—part I: Design theory, *IEEE Transactions on Microwave Theory and Techniques*, vol. MTT-20, no. 5, pp. 305–308, May 1972. xvii, 205, 206, 208, 209, 216

[35] D. J. Masse and R. A. Pucel, Microstrip propagation in magnetic substrates—part II: Experiment, *IEEE Transactions on Microwave Theory and Techniques*, vol. MTT-20, no. 5, pp. 309–313, May 1972. 205, 206, 208, 209, 210, 211

[36] R. F. Harrington, *Time Harmonic Electromagnetic Fields*, McGraw-Hill, NY, 1961. 205

[37] C. A. Balanis, *Advanced Engineering Electromagnetics*, 2nd ed., Wiley, NY, 2012. 205

[38] T. Kaneki, Analysis of linear microstrip using an arbitrary ferromagnetic substance as the substrate, *Electronics Letters*, vol. 5, no. 19, pp. 463–465, September 1969. 206

[39] R. A. Pucel, D. J. Masse, and C. P. Hartwig, Losses in microstrip, *IEEE Transactions on Microwave Theory and Techniques*, vol. MTT-16, no. 6, pp. 342–350, June 1968. 208, 209, 210, 221

[40] J. Green, E. Schlomann, F. Sandy, and J. Saunders, Characterization of the microwave tensor permeability of partially magnetized materials, *Semi-Annual Report RADC-TR-69–93*, February 1969. 210

[41] J. Green, F. Sandy, and C. Patton, Microwave properties of partially magnetized ferrites, *Rome Air Development Center*, Final Rep. RADC-TR-68–312, Rome, NY, August 1968. 210

[42] F. Sandy and J. Sage, Use of finite difference approximations to partial difference equations for problems having boundaries at infinity, *IEEE Transactions on Microwave Theory and Techniques*, vol. MTT-19, pp. 485–486, May 1971. 211

[43] N. G. Alexopoulos, Integrated-circuit structures on anisotropic substrates, *IEEE Transactions on Microwave Theory and Techniques*, vol. MTT-33, no. 10, pp. 847–881, October 1985. xvii, 212, 213, 214, 215

[44] R. P. Owens, J. E. Aitken, and T. C. Edwards, Quasi-static characteristics of microstrip on an anisotropic sapphire substrate, *IEEE Transactions on Microwave Theory and Techniques*, vol. MTT-24, pp. 499–505, August 1976. 212

[45] N. G. Alexopoulos and S. A. Maas, Characteristics of microstrip directional couplers on anisotropic substrates, *IEEE Transactions on Microwave Theory and Techniques*, vol. MTT-30, pp. 1267–1270, August 1982. 213

[46] L. Yu and B. Rawat, Quasi-static analysis of three-line microstrip symmetrical coupler on anisotropic substrates, *IEEE Transactions on Microwave Theory and Techniques*, vol. MTT-39, no. 8, pp. 1433–1437, August 1991. 213

[47] L. D. Landau and E. M. Lifshitz, *Electrodynamics of Continuous Media*, Addison-Wesley, Boston, 1960. 214

[48] B. T. Szentkuti, Simple analysis of anisotropic lines by a transform method, *Electronics Letters*, vol. 25, no. 12, pp. 672–673, December 9, 1976. 214

[49] T. C. Edwards and R. P. Owens, 2–18 GHz dispersion measurements of 10–100 Ω microstrip lines on sapphire substrate, *IEEE Transactions on Microwave Theory and Techniques*, vol. MTT-24, pp. 506-513, August 1976. 214

[50] M. Kobayashi, Frequency dependent characteristics of microstrips on anisotropic substrates, *IEEE Transactions on Microwave Theory and Techniques*, vol. MTT-30, pp. 2055–2057, November 1982. 214

[51] M. V. Schneider, Microstrip lines for microwave integrated circuits on anisotropic substrates, *Bell System Technical Journal*, vol. 48, pp. 1421–1444, May/June 1969. 215

[52] E. Yamashita, K. Atsuki, and T. Ueda, An approximate dispersion formula of microstrip lines for computer-aided design of microwave integrated circuits, *IEEE Transactions on Microwave Theory and Techniques,* vol. MTT-27, pp. 1036–1038, December 1979. 215

[53] S. Bolioli, H. Benzina, H. Baudrand, and B. Chan, Centimeter-wave microstrip phase shifter on a ferrite-dielectric substrate, *IEEE Transactions on Microwave Theory and Techniques,* vol. MTT-37, pp. 698–705, April 1989. xvii, 215, 216, 217, 218, 219

[54] D. Polder, On the theory of ferromagnetic resonance, *Philosophy Magazine,* vol. 40, no. 300, pp. 99–115, 1949. 220

[55] E. D. Palik and J. K. Furdyna, Infrared and microwave magnetoplasma effects in semiconductors, *Reports on Progress in Physics,* vol. 33, pp. 1191–1317, 1970. 220

[56] I. Burneuka, L. Knishevskaya, O. Smertin, M. Tamoshiunene, and V. Shugurov, Analysis of waveguide structures by means of singular integral equations, *International Journal of Infrared and Millimeter Waves,* vol. 17, no. 1, pp. 159–183, 1996. 220, 221

[57] L. Yu and B. Rawat, Quasi-static analysis of three-line microstrip symmetrical coupler on anisotropic substrates, *IEEE Transactions on Microwave Theory and Techniques,* vol. MTT-39, no. 8, pp. 1433–1437. 220

Authors' Biographies

JOHN N. SAHALOS

John N. Sahalos received B.Sc. degree in Physics, M.Sc. degree in Electronic Physics, BCE & MCE degrees in Civil Engineering, and Ph.D. in Electronic Physics from the Aristotle University of Thessaloniki (AUTH), Greece. Prof. Sahalos is with the Radio-Communications Laboratory at the AUTH, Greece and with the Department of Engineering at the University of Nicosia, Cyprus. For eight years he was a Professor in the ECE Department, University of Thrace, Greece, and Director at the Microwaves Laboratory. He was a visiting faculty member at Ohio State University, the University of Colorado, and the Technical University of Madrid. He was also on the board of directors of the National Research & Technology Committee of Greece and on the Board of Directors of OTE S.A., the largest telecommunications company in Southeast Europe. For more than 10 years, Prof. Sahalos was the president of the Greek committees of URSI. Also, he was a member of the committee and internal auditor of EurAAP (European Association of Antennas & Propagation). Besides his academic duties, he is now a Coordinator of the R&D advisory board at a high tech industry.

Prof. Sahalos is an IEEE Life Fellow, an Honorary Fellow of the Electronic Physics Society, a Fellow of the Physical Society, and a member of the Technical Chamber of Greece.

He is the author of four books, one in English (Wiley), and of more than 450 articles published in scientific literature. His research interests include antennas, radio communications, EMC/EMI, RFIDs, Microwaves, and Biomedical Engineering. With his colleagues, he designed innovative products like the EIT, the MLS, the ORAMA simulator and the SMS-K monitoring system.

GEORGE A. KYRIACOU

George A. Kyriacou was born in Famagusta, Cyprus, on March 25, 1959. He received his Electrical Engineering diploma and Ph.D. degree, both with honors, from the Democritus University of Thrace, Xanthi, Greece, in 1984 and 1988, respectively. Since January 1990, he has been continuously with the Department of Electrical and Computer Engineering, Democritus University of Thrace, where he is currently Professor and Director of the Microwaves Laboratory. From 2005 to 2010 he has been Director of the Graduate Studies of the department. He has authored over 250 journal and conference papers and supervised 7 Ph.D. and 11 M.Sc. theses and more than 90 Diploma theses. His main research interests include microwave engineering, open waveguides and antennas in anisotropic media, software-defined and cognitive radio, computational electromagnetics, and biomedical engineering. Dr. Kyriacou is member of IEEE since 1990 and senior member since 2000. He is also a member of the Technical Chamber of Greece and the European Microwave Association (EuMA).